Advances in Intelligent Systems and Computing

Volume 667

Series editor

Janusz Kacprzyk, Polish Academy of Sciences, Warsaw, Poland
e-mail: kacprzyk@ibspan.waw.pl

The series "Advances in Intelligent Systems and Computing" contains publications on theory, applications, and design methods of Intelligent Systems and Intelligent Computing. Virtually all disciplines such as engineering, natural sciences, computer and information science, ICT, economics, business, e-commerce, environment, healthcare, life science are covered. The list of topics spans all the areas of modern intelligent systems and computing such as: computational intelligence, soft computing including neural networks, fuzzy systems, evolutionary computing and the fusion of these paradigms, social intelligence, ambient intelligence, computational neuroscience, artificial life, virtual worlds and society, cognitive science and systems, Perception and Vision, DNA and immune based systems, self-organizing and adaptive systems, e-Learning and teaching, human-centered and human-centric computing, recommender systems, intelligent control, robotics and mechatronics including human-machine teaming, knowledge-based paradigms, learning paradigms, machine ethics, intelligent data analysis, knowledge management, intelligent agents, intelligent decision making and support, intelligent network security, trust management, interactive entertainment, Web intelligence and multimedia.

The publications within "Advances in Intelligent Systems and Computing" are primarily proceedings of important conferences, symposia and congresses. They cover significant recent developments in the field, both of a foundational and applicable character. An important characteristic feature of the series is the short publication time and world-wide distribution. This permits a rapid and broad dissemination of research results.

More information about this series at http://www.springer.com/series/11156

Rituparna Chaki · Agostino Cortesi
Khalid Saeed · Nabendu Chaki
Editors

Advanced Computing and Systems for Security

Volume Six

 Springer

Editors

Rituparna Chaki
The A.K. Choudhury School
  of Information Technology
University of Calcutta
Kolkata, West Bengal
India

Agostino Cortesi
Computer Science DAIS
Università Ca' Foscari
Venice
Italy

Khalid Saeed
Faculty of Computer Science
Bialystok University of Technology
Białystok
Poland

Nabendu Chaki
Department of Computer Science
  and Engineering
University of Calcutta
Kolkata, West Bengal
India

ISSN 2194-5357   ISSN 2194-5365 (electronic)
Advances in Intelligent Systems and Computing
ISBN 978-981-10-8182-8   ISBN 978-981-10-8183-5 (eBook)
https://doi.org/10.1007/978-981-10-8183-5

Library of Congress Control Number: 2017964593

Printed on acid-free paper

This Springer imprint is published by the registered company Springer Nature Singapore Pte Ltd. part of Springer Nature
The registered company address is: 152 Beach Road, #21-01/04 Gateway East, Singapore 189721, Singapore

Preface

The Fourth International Doctoral Symposium on Applied Computation and Security Systems (ACSS 2017) took place on March 17–19, 2017, in Patna, India.

The University of Calcutta along with Ca' Foscari University of Venice, Bialystok University of Technology, and Warsaw University of Technology collaborated to make ACSS 2017 a grand success. Around 40 participants from a multitude of institutions have taken part in a highly interactive discussion for 3 days, resulting in a cumulative experience of research idea exchange.

The post-conference book series are indexed by ISI Web of Sciences. The sincere effort of the program committee members coupled with ISI indexing has drawn a large number of high-quality submissions from scholars all over India and abroad. The Technical Program Committee for the symposium selected only 21 papers for publication out of 70 submissions.

The papers mainly covered the domains of computer vision and signal processing, biometrics-based authentication, security for Internet of things, analysis and verification techniques, security in mobile and cloud scenarios, large-scale networking, remote health care, distributed systems, software engineering, cloud computing, privacy and confidentiality, access control, big data and data mining, android security.

The technical program was organized into six topical sessions each day. The sessions started with a keynote lecture on a pertinent research issue by an eminent researcher/scientist. This was followed by short, to-the-point presentations of the technical contributions. At the end of each session, the session chair handed over the suggestions for improvement pertaining to each paper. The sessions also saw lively discussions among the members of the audience and the presenters.

The post-conference book includes the presented papers in enhanced forms, based on the suggestions of the session chairs and the discussions following the presentations. Each of the accepted papers had undergone a double-blind review process. During the presentation, every presented paper was evaluated by the concerned session chair, an expert in the related domain. As a result of this process, most of the papers were thoroughly revised and improved, so much so that we feel that this book has become much more than a mere post-workshop proceedings volume.

We would like to take this opportunity to thank all the members of the Technical Program Committee and the external reviewers for their excellent and time-bound review works. We thank all the sponsors who have come forward toward the organization of this symposium. These include Tata Consultancy Services (TCS), Springer India, ACM India. We appreciate the initiative and support from Mr. Aninda Bose and his colleagues in Springer for their strong support toward publishing this post-symposium book in the series "Advances in Intelligent Systems and Computing." Last but not least, we thank all the authors without whom the symposium would not have reached up to this standard.

On behalf of the editorial team of ACSS 2017, we sincerely hope that this book will be beneficial to all its readers and motivate them toward further research.

Kolkata, India   Rituparna Chaki
Venice, Italy   Agostino Cortesi
Białystok, Poland   Khalid Saeed
Kolkata, India   Nabendu Chaki

Contents

About the Editors

Rituparna Chaki is Professor of Information Technology at the University of Calcutta, India. She received her Ph.D. from Jadavpur University in India in 2003. Before this, she completed her B. Tech. and M. Tech. in Computer Science and Engineering from the University of Calcutta in 1995 and 1997, respectively. She has served as a System Executive in the Ministry of Steel, Government of India, for 9 years, before joining the academics in 2005 as a Reader of Computer Science and Engineering in the West Bengal University of Technology, India. She is with the University of Calcutta since 2013. Her areas of research include optical networks, sensor networks, mobile ad hoc networks, Internet of things, data mining, etc. She has nearly 100 publications to her credit. She has also served in the program committees of different international conferences. She has been a regular Visiting Professor at the AGH University of Science and Technology, Poland, for last few years. She has co-authored a couple of books published by CRC Press, USA.

Agostino Cortesi, Ph.D., is a Full Professor of Computer Science at Ca' Foscari University, Venice, Italy. He served as Dean of the Computer Science Studies, as Department Chair, and as Vice-Rector for quality assessment and institutional affairs. His main research interests concern programming languages theory, software engineering, and static analysis techniques, with particular emphasis on security applications. He has published more than 110 papers in high-level international journals and proceedings of international conferences. His h-index is 16 according to Scopus and 24 according to Google Scholar. He has served several times as member (or chair) of program committees of international conferences (e.g., SAS, VMCAI, CSF, CISIM, ACM SAC), and he is in the editorial boards of the journals "Computer Languages, Systems and Structures" and "Journal of Universal Computer Science." Currently, he holds the chairs of "Software Engineering," "Program Analysis and Verification," "Computer Networks and Information Systems," and "Data Programming."

Khalid Saeed is a Full Professor in the Faculty of Computer Science, Bialystok University of Technology, Bialystok, Poland. He received his B.Sc. in Electrical and Electronics Engineering from Baghdad University in 1976 and M.Sc. and Ph.D. from Wroclaw University of Technology in Poland in 1978 and 1981, respectively. He received his D.Sc. (Habilitation) in Computer Science from Polish Academy of Sciences in Warsaw in 2007. He was a Visiting Professor of Computer Science at Bialystok University of Technology, where he is now working as a Full Professor. He was with AGH University of Science and Technology in 2008–2014. He is also working as a Professor in the Faculty of Mathematics and Information Sciences at Warsaw University of Technology. His areas of interest are biometrics, image analysis and processing, and computer information systems. He has published more than 220 publications and edited 28 books, journals and conference proceedings, 10 text and reference books. He has supervised more than 130 M.Sc. and 16 Ph.D. theses. He gave more than 40 invited lectures and keynotes in different conferences and at universities in Europe, China, India, South Korea, and Japan on biometrics, image analysis and processing. He has received more than 20 academic awards. He is a member of more than 20 editorial boards of international journals and conferences. He is an IEEE Senior Member and has been selected as IEEE Distinguished Speaker for 2011–2016. He is the Editor-in-Chief of International Journal of Biometrics with Inderscience Publishers.

Nabendu Chaki is a Professor in the Department of Computer Science and Engineering, University of Calcutta, Kolkata, India. He did his first graduation in Physics from the legendary Presidency College in Kolkata and then in Computer Science and Engineering from the University of Calcutta. He has completed Ph.D. in 2000 from Jadavpur University, India. He is sharing six international patents including four US patents with his students. He has been quite active in developing international standards for Software Engineering and Cloud Computing as a member of Global Directory (GD) for ISO-IEC. Besides editing more than 25 book volumes, he has authored 6 text and research books and has more than 150 Scopus-indexed research papers in journals and international conferences. His areas of research interests include distributed systems, image processing, and software engineering. He has served as a Research Faculty in the Ph.D. program in Software Engineering in US Naval Postgraduate School, Monterey, CA. He is a visiting faculty member for many universities in India and abroad. Besides being in the editorial board for several international journals, he has also served in the committees of over 50 international conferences. He is the founder Chair of ACM Professional Chapter in Kolkata.

Part I
Algorithms

A Framework for Solution to Nurse Assignment Problem in Health Care with Variable Demand

Paramita Sarkar, Ditipriya Sinha and Rituparna Chaki

Abstract The goal of this research is to evolve nurse scheduling problem as a matured computational model towards optimization of multiple and conflicting objectives, the complex shift patterns, etc., in changing the environment. This work aims to find out an effective assignment of nurses to home patients as well as in-hospital patients based on patients' varying demands over time according to their health status. It proposes nurse scheduling algorithms based on variable time quantum, wait time, context switch time, etc., in different situations when the environment becomes more constrained as well as unconstrained. It develops corresponding cost functions for assigning suitable nurses by considering the penalty cost, swapping cost of nurses. This paper proposes methods to utilize nurses by using nearest neighbour based similarity measure and combined genetic algorithm to generate feasible solutions. Finally, this paper implements the proposed algorithms and compares these with the other popular existing algorithms.

Keywords Nurses scheduling · Home care · Hospital care · Variable time quantum · Genetic algorithm · Cost · Nearest neighbour

1 Introduction

This research work investigates the well-researched nurse scheduling problem with proposing a variable time quantum based nurse assignment. It opens up various problems and solutions in assigning nurses to patients in required shift patterns on the

P. Sarkar (✉)
Calcutta Institute of Engineering and Management, Tollygunge, Kolkata, India
e-mail: mailtoparo@gmail.com

D. Sinha
National Institute of Technology Patna (NIT-PATNA), Patna, India
e-mail: ditipriyasinha87@gmail.com

R. Chaki
A. K. Choudhury School of Information Technology, University of Calcutta, Kolkata, India
e-mail: rituchaki@gmail.com

© Springer Nature Singapore Pte Ltd. 2018
R. Chaki et al. (eds.), *Advanced Computing and Systems for Security*, Advances in Intelligent Systems and Computing 667,
https://doi.org/10.1007/978-981-10-8183-5_1

basis of their appropriate expertise and grade with severity of the patients in any hospital, nursing centre or in remote healthcare services every day. Researchers have been working with various mathematical descriptions and techniques [1–5] to solve this. Here, we have considered the domain of remote health care, where patients at remote places may need continuous monitoring services and advices from the caregivers. In emergency situation, the critical and remote patients need most prioritized services than other normal patients; therefore some efficient techniques need to be designed to provide appropriate nursing. In some situations, the demand of the patients to be assigned is varying according to the urgency of the environmental situation. It involves staffing (how many nurses are needed, when, and where) and scheduling (which determines when and where each nurse works), penalty cost (number of violations of the hospital restrictions). These are part of an interrelated, hierarchical problem. Once a staffing plan is finalized, a schedule is constructed that informs individual nurses and nurse head of who is assigned when and where. This research considers only the scheduling aspect of the larger problem. This research work implements the proposed algorithm nurse assignment procedure in remote centre (VTQNS) in remote healthcare environment where every nurse is assigned to remote patient from the available nurse pool of the hospital. Every nurse in the remote centre is scheduled to serve at the patient's place in a variable time period where the fixed time duration is preset. Then it is modified according to the status of the patient, patient's status is checked at regular time interval. Therefore, nurses are swapped when and as needed as per the nurses' grades and patient's condition.

2   Review Work

This section highlights the significant relevance of this research work on nurse scheduling of various demand. The review section mainly discusses on how the patients are assigned to the appropriate nurses based on the priority, the techniques with which the nurses are scheduled, heuristic methods like evolutionary algorithms, hybrid heuristic methods, and local search techniques that have been used in several works on nurse scheduling.

2.1   Heuristic Search

Other than simple genetic algorithm, several local search techniques were applied in nurse scheduling problem earlier. In [6], a heuristic method for nurse scheduling is proposed to fulfil the real-life hospital limitations and nurse preferences while minimizing the patients' waiting queue time. The nurse rostering problem is solved in [7] by applying an adaptive heuristic method in the construction of the nurse schedule. In this paper, nurses are ordered and selected adaptively according to the quality of the schedules they were assigned to in the last iteration. It shows that the local search heuristic during the roster construction can further improve the constructed solutions for the benchmark problems tested. Meat heuristic is applied [8] to present and simulate the attraction and repulsion between shift patterns of nurses according to Coulomb's Law. This approach can be an important participant in the hybridization of heuristic procedures for the nurse scheduling problem. Another paper [9] proposes an automated nurse scheduling approach in two stages. First, an efficient data warehouse system based online analytical method is proposed for hospital information system. Subsequently, Enhanced Greedy Optimization algorithm is implemented to optimize the nurse roster and compared with other optimization algorithms. Other mathematical programming approaches are applied to develop heuristic [10] for generating feasible working shift patterns in a complex task oriented and uncertain environment. To fulfil the business requirements, there are few other constraints that have to be added with current proposed solution. In [11], an effective late acceptance hill climbing method is proposed to reduce the complexity of the real-time timetable problem.

Cost-Based Heuristic

In nurse scheduling problem, cost is a dominant parameter to build multi-objective function and to reduce the search space in the robust optimized solution. Aversion of nurses and the hospital cost could be dominant parameters to formulate objective function. A flexible nurse scheduling model is developed by the implementation of a mixed integer programming model in [10]. It optimizes both hospital's requirement as well as nurse preferences by allowing flexibility in the transfer of nurses from different duties. Different approaches are available with a cost matrix based simulated annealing (CMSA) in [12] to find a schedule that optimizes a set of constraints. They aim to minimize the cost function which is formulated here by assigning weights. The proposed simulated annealing provides an acceptable good solution in a fixed amount of time for transition rule, rather than an optimal solution. To reduce a large space complexity, paper [13] implements a method of coding using actual shifts and constraints for NSP. This protocol reduces the searching area by exchanging number of shifts between different groups of nurses. Penalty cost is often used in constrained optimization problem. But the experiment on nurse rostering problems demonstrates that the stochastic ranking [14] method is better in finding feasible solutions but fails to obtain good results with regard to the objective function in [15].

Genetic Algorithm

Genetic algorithms have been successfully used for nurse scheduling problems in the past and they always had to overcome the limitations of the classical genetic algorithms paradigm in handling the conflict between objectives and constraints. Genetic algorithm approach [16] is described to manpower scheduling problem in a UK-based hospital. It uses an indirect coding based on permutations of the nurses, and a heuristic decoder that builds schedules from this permutation, but this restricts the use of exact methods that search for global optima. Therefore, constructive heuristics [17] is applied based on specific encoding and operators for sequencing problems to nurse re-rostering problem under hard constraints. The complexity of this problem led the authors to design heuristic procedures. Another paper [18] proposes a strategy to improve performance of GA for NSP by using cost bit matrix for genetic operators, where the mutation operator was applied based on the value in the cost bit matrix also. Thus, search space and execution time are reduced. A two-stage mathematical modelling [19] is developed for a nurse scheduling system which shows that different infeasibilities of genetic algorithm approaches in nurse scheduling problems are used to find a better feasible solution [20]. A combined genetic algorithm with hyper simulated annealing is tested for stochastic ranking method [14] of nurse scheduling and provides higher performance than GA with penalty cost. But the premature convergence of traditional genetic algorithm is solved by using memetic algorithm [21, 22]. A combined local search based algorithm hill climbing is incorporated in genetic algorithm to result in better performance of NP-hard nurse scheduling problem. In [5], the addition of the hill climbing operator after each mutation operator greatly increases the speed at which better solutions are found above the evolutionary operators alone whereas violation directed hierarchical hill climbing [11] is executed with GA applying various crossover operators to compare among them.

3   Research Scope

As observed from the study of existing works in this area, we have noted the research gaps as follows:

- Existing works have not considered real-time dynamic decision-making to identify the appropriate nurses for replacement or shuffling according to the priority level of the patients.
- The existing works tend to have problems in deciding upon the multi-objective function considering soft and hard constraints, only in specific time period and for specific use case.
- There is lack of discussion of nurse assignment in emergency situation.
- Existing works did not quantify the amount of duty hours to which extent a nurse is utilized in periodic assignment.

4 Proposed Work

4.1 Framework

Keeping in mind the above research gaps, the following framework for nurse assignment has been proposed (Fig. 1).

The proposed work has tried to shuffle the schedule of nurses based on the variable time quantum required to assign nurses to higher priority patients.

This paper proposes scheduling algorithms for solving utilization problem in both home care and hospital care nursing services. The proposed framework consists of three main modules. They are central nursing care service module, home care nursing service (Nurse scheduling for home patient assignment), and hospital care nursing service (Nurse scheduling for hospital patient assignment).

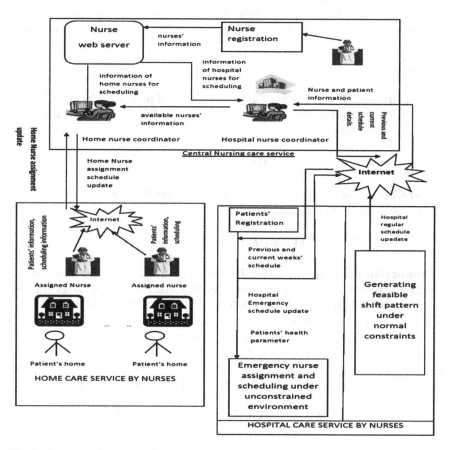

Fig. 1 Framework for nurse assignment with varying demand

4.2 Description of Proposed Nurse Assignment Methods

The proposed nurse assignment method for variable demand has been conceptualized in three different aspects which are discussed as follows.

Central Nursing Care Service Module

The aim of this module is to generate and maintain a master nurse-base. Nurse web server is used to store all information after the nurse's registration along with other scheduling information which are transferred in-between all sub-modules.

Home Care Nursing Service

This module is designed to provide nursing service to patients at their respective homes. The request as per patient's severity is communicated to central nursing care service module. These schedules are updated as per the proposed variable time quantum nurse scheduling (VTQNS) algorithm which is designed based on variable request time and the response time for different patients according to their priorities (Table 1).

Definition

Time quantum (TQ): It is defined as the variable total time which is evaluated as total duty hours T_i served by a nurse plus waiting time WT for each patient who is assigned to that nurse.

$$TQ = T_i + WT \tag{1}$$

Turnaround Time (TAT): It is the overall time required for a nurse to be assigned to a remote patient, as given in Eq. (2).

$$TAT = \left(\text{Start Time } (t_{sa}) - \text{Request time } (t_{req}) + CS\right) \tag{2}$$

Traffic state function $f(j, P[k])$: The state of the traffic on the high road is measured by the probability of occurrences of the traffic jam at four different times in a day such as EARLY MORNING(0), MID DAY(1), AFTERNOON(2), LATE NIGHT(3). The state of the traffic is expressed as a function $f(j, P[k])$ of jth shift and traffic probability, where $P[k]$ is the probability of occurrences of k number of vehicles for traffic jam in the high road, as $0.5 < P[k] \leq 1$. Following the Poisson's distribution, G is the mean number of traffic per unit time. If $J \in D$ where $D = \{0, 1, 2, 3\}$ then, when $j \in J \{0, 3\}$, $P[k] < 0.5$, and when $j \in J \{1, 2\}$, $0.5 < P[k] \leq 1$.

$$
\begin{aligned}
f(j, P[k]) &= +1 \text{ if } j \in J \{0, 3\} \text{ and } P[k] < 0.5 \\
&= 0, \text{ if } j \in J \{1, 2\} \text{ and } 0.5 \leq P[k] \leq 1
\end{aligned} \tag{3}
$$

Table 1 Index table

Index	Meaning
TQ_i	Variable time quantum of a nurse for a patient i
N	Total number of nurses
P_{ass}	Total number of assigned patient
AVAN[N]	Available pool of free nurses
RQ(P, s)	Ready queue of remote patients (P), distances (s) from centre
N_i	ith nurse deleted from the available pool
T_i	Scheduled duty hours of ith nurse
STATUS[j]	Array of the status of jth home patient
START$_p$	Start time of patient p
WT	Waiting time (Start Time-Arrival Time)
PR$_t$	Priority of a patient at any time instant t
ET	Effective time of assignment
TAT	Turnaround Time
$f(j, P[k])$	Traffic state function
$t_{sa}(P_{i+1})$	The start time for a patient P_{i+1}
t_{req}	Request time of a patient P_i
$t_{free}(P_i)$	The finish time for a patient P_i
T_{Total}	Total time required by each assigned nurse
t_{n1Pj}	Time to over the distance by a nurse $n1$ to reach the place of a patient P_i
a_{ip}	Assignment factor of ith nurse for pth patient
k_i	The scheduled service hours for each assigned patient i

Context Switch (CS): It is the time required by a nurse $n1$ to switch from one patient P_i to another patient P_{i+1}. Patient P_{i+1} immediately follows patient P_i including the travelling time $t_{n1P_{i+1}}$ and the handover time t_{hanP_i} from the previous nurse to the new nurse.

$$CS = \left[\left(t_{n1P_{i+1}} + t_{handi} \right) + \left(t_{sa}(P_{i+1}) - t_{free}(P_i) \right) \right], \tag{4}$$

where t_{hand} is the time taken by the nurse to hand over the charge to another nurse at the time of released from the duty.

$$t_{n1P_j} = s/v \times f(j, P[k]) \tag{5}$$

Algorithm: VTQNS(RQ[NUM], AVANP[N], ARRIVALp, t$_{sa}$, t$_{req}$, TQ, TAT).
Input: RQ[NUM], AVANP[N], ARRIVALp, t$_{sa}$, t$_{req}$
Output: TQ, TAT.

ALGORITHM VTQNS

1. sort (RQ, NUM) and sort (AVANP,N) in descending order according to priority and grades respectively.
2. For each nurse i→1 to N
       Delete i from the AVANP [N]
3. For a patient j → 1 to NUM repeat steps 3.1 to 3.9
3.1 Assignment of N_i to single patient j from RQ, Set N_i → P_j
3.2 [Initialization of time quantum] TQ=1 for each N_i,
and set, TOT_PRI=0
3.3 *while* (TQ≤T_i) calculation of effective time ET of each assigned nurse N_i
3.4 Set, ET= WT+ (t_{free} − t_{sa})
3.5 [Storing the status of priority PR of current assigned patient]
3.6 if priority does not changes at time instant t+1,
    then, ΔPR ≠ 0, Set Set, STATUS_N_i[j]=PR_t.
    otherwise
                PR_{t+1} is stored STATUS[j]=PR_j±ΔPR
3.7 [Updation of total priority value of all assigned patients to N_i]
Set, TOT_PRI=TOT_PRI + STATUS_N_i [NUM]
3.8 [Evaluation of average priority]
Set, AVG_PRI=(TOT_PRI)/NUM.
3.9 if AVG_PRI>=3,
        then update time quantum as TQ=TQ+1 and go to step 7
        Else
            I.   delete patient from RQ
            II.   if TQ (N_i)=T_i then go to step 6
4. Updation of nurse index i and patient index j and go to step 7
       End of step 2 loop.
5. Set, TQ=TQ+ET.
6. Calculate total number of remote patients n_pat assigned to each nurse N_i
7. Evaluation of TAT, T_{total} for each nurse.

Fig. 2 Proposed VTQNS algorithm

Assumptions

- Each patient is registered with the nearest nursing centre. Each nurse is assigned at most for maximum 6-h duration in each day.
- Each nurse can serve at least three consecutive night (*N*) shifts in the preceding week.
- At most four home patients can be assigned to an individual nurse in a week (Fig. 2).

Hospital Care Nursing Service

The following subsections have described on the nurse scheduling algorithms in two situations inside the hospital. Patients are registered in patients' registration sub-module with information like patients' name, age, gender, health_stat. The other sub-module is designed for unconstrained emergency situation in hospital based

Table 2 Symbol definition table for hospital care nursing service

Symbol	Definition
C_{TOT}	The fitness function of combined GA
$F(s)$	The set of feasible schedule pattern
X_{is}	$X_{is} = 1$ for each feasible shift pattern of a nurse
t_{ie}	ith nurse's grade index with grade e
R_{ijk}	$R_{ijk} = 1$ if ith nurse has preference for jth shift of kth day, else 0
$\varnothing(x)$	The transformed function
$g_j(x)$	Violation function measures the total number of violation (x) in a shift j
ω_g	The associated penalty coefficient
λ	The Poisson distribution parameter
a_{ijk}	Assignment variable $a_{ijk} = 1$, if a nurse is assigned in jth shift on kth day
$\omega_{i'}$	Weight parameter of the swapped nurse
prod	The productivity of the assignment of available nurses to the patients in wait queue

on the previous week's schedule and depends on the availability of the appropriate nurses. Another one is for regular nurse scheduling considering various hard and soft constraints. In this module, the nurses are assigned according to the schedules developed by the hospital nurse coordinator (Table 2).

Nurse Scheduling in Constraint Environment

In this section, an algorithm for nurse scheduling under constrained situation is described. Combined genetic algorithm (GA) with hill climbing (HC) method is implemented here to generate feasible schedule for 1 week under normal working constraints. Inside the hospital, the feasibility is obtained by applying the satisfying all constraints imposed by the hospital's rules. GA and local search help in optimization of objective function with fast convergence. In this paper, some hard constraints are assumed as follows:

- No nurse is assigned to work a night shift followed immediately by a day shift.
- Higher priority patients should replace lower priority patients in each shift type.
- Minimum number of assigned patient in one shift type is one.
- Preference of a higher graded nurse is higher than lower graded nurse in all shift types.
- Highest graded nurse should be assigned to highest priority patient in all shift types.

The fitness function (C_{TOT}) in this combined algorithm is used as the following cost function.

$$C_{\text{TOT}} = \sum_{i=1}^{n} \sum_{j=1}^{2} p_{ij} X_{is} + T_{\text{wait}}$$

$$+ \sum_{s \in F(i)} \sum_{k=1}^{7} \min \left[\sum_{i=1}^{n} \sum_{j=1}^{2} |(X_{is} t_{ie} a_{ijk}) - R_{ijk}| ; 0 \right] \qquad (6)$$

Algorithm for combined genetic algorithm-Combined_GA(F(s),Gensize,S).
Input: F(s), Gensize
Output: The best solution S and fitness function C_{TOT}
This algorithm generates the best solution S by implementing the fitness function C_{TOT} on each generation Gen.

```
Steps:
1. Generate solution space as feasible schedule pat-
tern set F
2. Set Gen=0, Initialize population P randomly

4. For each individual n∈ P: calculate fitness(n) as
C_TOT as in equation (6);
5. While (Gen < Gensize) Apply generic GA with se-
lection(n), cross-over(n), mutation(n);

   for each individual n∈ P, do algorithm HillClimb-
ing_search(n, fitness)
        do replacement(n)
6. If the termination condition is satisfied, stop
else return the best solution S in current popula-
tion and go to Step 5.
```

In this Hillclimbing_search, if the newly generated chromosomes have better fitness then it replaces the old chromosome else checks the loop condition.

Emergency Nurse Scheduling Inside the Unconstrained Hospital Environment

Efficient Nurse scheduling algorithm is required in healthcare systems for faster assignment in case of emergency situation. The design of the proposed assignment process is discussed herein two possible situations. In this paper, the nurses are fetched in the emergency situation from the available pool of nurses who have not assigned on that day or in leave. The schedule of the previous week is required here to utilize the uncovered shift patterns of which day or of which nurses. Higher priority nurses are assigned according to their grades in unconstrained environment. The minimum need of a nurse in emergency case is considered to be assigned for at least minimum duty hours (Mhr) in the emergence period. The following subsection discusses on how penalty function and similarity functions are used in unconstrained scheduling in this paper.

Duty level penalty cost. A penalty function $\varphi_1\left(g_j\left(x\right)\right)$ is used here for measuring the total number of violation (x) of minimum need of feasibility in case of emergency situation.

Schedule pattern penalty cost. On previous day, for any uncovered shift pattern, if the corresponding schedule is infeasible then the nurse is penalized for that given schedule. That penalty function is given as the maximum of nurse's preferences in a week:

$$\varphi_2(g_j(x)) = \sum_{k=1}^{7} \sum_{j=1}^{J} \max \left[0, R_{kr} \times \sum_{k=1}^{7} \sum_{j=1}^{J} a_{ijk} \right], \tag{7}$$

where for $a_{ij'k'} X_{is} = 0$, $j' \in \{J - j\} k' \in \{K - k\}$, if $g_j(x) : a_{ijk} X_{is} = 1$

Those nurses with high penalty value will be stored in the previous Max Priority Queue for assigning them with high priority also. The transformed function $\varnothing(x)$ is designed here to convert the cost function in Eq. (6) to unconstrained objective function in Eq. (8) which is represented as

$$\varnothing(x) = C_{TOT} + \lambda \omega_g \left(\varphi_1 \left(g_j(x) \right) + \varphi_2 \left(g_j(x) \right) \right) \varnothing > 0 g_j(x) \geq 0 \tag{8}$$

Distance-Based Similarity Measure

In our paper, ωg is used as the associated penalty coefficient which is the distance based on nearest neighbour joining between two shift patterns of two distinct nurses with a phylogenetic tree [23]. The objective of this section is to evaluate the similar efficiency of the nurses so that the 'lesser' assigned nurses can be utilized in the next week's schedule and if they are not available then their closely related nurse can be substituted. Under emergency condition, the assignment situation is considered unconstrained, no hospital restriction is imposed on the scheduling. In this method, two solutions of unconstraint situation are compared based on the nearest neighbour joining measure [23] which is a suboptimal solution to build the topology as well as the tree of all similar workload of nurse. In this matrix, we store the distance between two similar chromosomes. The distance between two different chromosomes is calculated as the total number of different 1 bit between them. There are examples of five distinct chromosomes (A, B, C, D, E) which represent the previous weekly schedule of individual nurse.

A: 11100111001110

B: 11101011111011

Neighbour joining takes as input a distance matrix specifying the distance between each pair of chromosome. The algorithm starts with a completely unresolved phylogenetic tree, it iterates over the following steps until the tree is completely resolved and all branch lengths are known. This is shown in the following matrix A (Fig. 3).

Neighbour joining procedure: The pairs of distinct chromosome i and j (i.e. with $i \neq j$) for which $Q(i, j)$ has its lowest value are joined. Based on a distance matrix relating the n chromosomes, Q is calculated as follows:

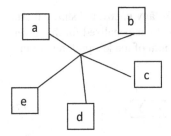

chromosomes	A	B	C	D	E
A	0	2	1	5	3
B	2	0	2	3	1
C	1	2	0	3	2
D	5	3	3	0	2
E	3	1	2	2	0

Fig. 3 Phylogenetic tree and matrix A

$$Q(i, j) = (n - 2) d(i, j) - \sum_{k=1}^{n} d(i, k) - \sum_{k=1}^{n} d(j, k) \tag{9}$$

These chromosomes of two shift pattern are joined to a newly created node, which is connected to the central node. The new distance from the new node u of D and E is calculated.

$$\delta(D, u) = \frac{1}{2} d(D, E) + \frac{1}{2(n-2)} \left[\sum_{k=1}^{n} d(D, k) - \sum_{k=1}^{n} d(E, k) \right], \tag{10}$$

where u is the new node, k is the node which we want to calculate the distance to and D and E are the members of the pair just joined. Now these a and b nurses can replace the nurse u of the previous week schedule. Finally, the closest grade of the nurse is chosen from the connected branches. Thus, penalty coefficient is obtained as follows:

$$\omega_g = \min(d(D, u), d(E, u)) + d_{avg} \text{ (All of the new nodes)} \tag{11}$$

5 Experimental Results and Discussion

5.1 Performance Evaluation of Home Care Nursing Service

In the home care nursing service module, we have proposed VTQNS in C language. We have measured the parameters in the following tables which comply with the existing algorithm like CMGA, Simulated Annealing [12]. All the constraints described in this paper were included in both methods with exact same conditions. The goal was to check whether the proposed algorithm could actually generate an acceptable priority-based NSP for all graded nurses and remote patients in lesser time. Thus, the proposed VTQNS algorithm has been implemented to analyse the time regarding the nurse scheduling for a good number of patients with real-time data.

Table 3 Comparison between VQTNS and GA, CMGA

N_id	Grade	CS	T_{total}	SW $(N_i, N_{i'})$	COST$_{\text{total}}$ (N_i)
2234	1	25.2	54	12.6	16.2
1567	1	11.17	30	0	8.2
1310	2	17	50	15	18

Calculation of Total Time T_{total} and COST$_{\text{total}}$ Taken by Each Nurse and Cost Function

Total time T_{total} is calculated for each nurse as the sum of the computed TQ and the higher value between the maximum duty hours (k) and the total scheduled time (T_i) taken for all patients.

$$T_{\text{total}} = \int \sum_{i=1}^{P} \text{TQ}_i + \left[\left(\max \left[\sum_{i=1}^{n} T_i, k_i \right] \right) \times \sum_{p} a_{ip} \right], \qquad (12)$$

where p = the total number of scheduled patients and

$$T_i = \left(t_{\text{free}} - t_{sa_i} + \text{TAT} \right) \leq k_i, \qquad (13)$$

$a_i = +1$, if she is assigned to the scheduled patient i from the list $a_i = -1$ otherwise. Therefore, total cost of each nurse N_i is expressed as

$$\text{COST}_{\text{total}} (N_i) = \left[\left(T_{\text{total}} \times W (N_i) \right) - \left(\sum_{p=1}^{N} \text{CS} (N_{iP}) + \text{SW} (N_i, N_{i'}) + \partial (N_{i'}) \right) \right] \qquad (14)$$

$$W (N_i) = +1, \text{ if } T_{\text{wait}} > 0.5, W (N_i) = 0, \text{ otherwise} \qquad (15)$$

SW $(N_i, N_{i'})$ is the swapping cost which is required if any nurse N_i serves more than 6 h a day to one patient. Then, an equivalent efficient nurse N_i' is searched among the priority queue with time $O(n \log n)$ provided the required time for nurse N_i', i.e. $t_i' < 6$ and she is replaced or swapped with the nurse N_i. Otherwise N_i cannot be swapped and $t_i + h$ hours is added to the calculated TQ.ai = assignment factor of the particular nurse and Pp = waiting factor of a nurse. W_p = priority index of assigned patient. It is +1 if corresponding priority PR$_t$ > AVG_PRI, otherwise it is -1. Weight parameter of the swapped nurse is measured as $\omega_{i'} = +1$, if $q_{i'} > q_i$, $\omega_{i'} = 0$ otherwise

$$\text{SW} (N_i, N_{i'}) = W_p \times \text{PR}_t \times \left(\omega_{i'} \times \left[(6 - t_{i'P1}) + \text{TQ}_{i'} \right] \right) \qquad (16)$$

$\partial (N_{i'})$ is the violation cost which is added when a nurse in the available nurse pool is violating the hospital constraints in the normal scheduling (Table 3).

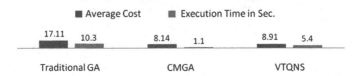

Fig. 4 Bar chart represents the comparison between VQTNS and GA, CMGA

Table 4 Table to show the result of average distance and time

Method	Similarity measurement	Time
Neighbour joining distance	Distance measure = 0.786	26.8
Set covering problem	Average all row cost rank = 2.8	33.6

This table shows the results generated for each nurse assigned to all assigned remote patients (Fig. 4).

This proposed work has been compared to CMGA [18], where average cost f_{avg} is 8.14 using GA with mutation probability 0.01. But in this VTQNS approach, priority- and grade-based approach is implemented considering variable waiting time, turnaround time, time quantum and swapping cost, violation cost for three patients for each nurse. This approach generated a nurse schedule faster in average cost than traditional GA, but the average cost is very near to the cost of CMGA in [5]. Although we have presented this work in terms of nurse scheduling, it should be noticed that the main idea of the approach could be applied to many other scheduling problems (Table 4).

5.2 Results of Hospital Care Nursing Service in Emergency Scheduling

In Table 2, the proposed nearest neighbour joining measure and score-based ranking using genetic algorithm in set covering problem [22] have been compared. In [22], different heuristic approaches were implemented and the results showed that too much extra calculation are needed for applying additional decoder and extra optimization layer to find improved solution.

Analysis of Graph of Productivity of Assignment Over Scheduled Duty Hours in Emergency Scheduling

The productivity of the assignment of available nurses to the patients in wait queue is measured by the ratio between the total number of assigned patient (Pass) and the total number of available nurses (AVAN[N]) at time t.

Fig. 5 Graphs of
productivity over time

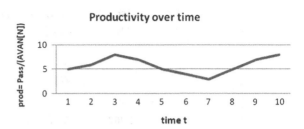

Table 5 Comparison between generic GA versus combined GA

Feasible solutions in for different populations	Generic GA	Computation time	Proposed combined GA	Computation time
No. of feasible solution	3	5.267	5	4.053
No. of feasible solution	0	3.21	1	2.33
No. of feasible solution	0	1.012	2	1
No. of feasible solution	2	3.424	3	4.21

$$\text{prod} := \frac{P_{\text{ass}}}{\text{AVAN}[N]} \tag{17}$$

As P_{ass} varies with time, so productivity in emergency situation will also vary as time. The above graph in Fig. 5 shows that at certain time when number of assigned patients becomes lesser than the number of available nurses in the pool, then the productivity factor decreases accordingly. To make the situation better, it requires to minimize the waiting time WT for every patient in the queue as well as to increase the number of assignment in emergency situation.

5.3 Results of Combined GA for Regular Scheduling

The proposed work in the regular scheduling module inside the hospital care nursing service has been implemented based on combined GA with local search in Turbo C. The following graphs show the results for both generic GA and combined GA.

Graphs of Feasible Solution and Computation Time of Generic GA versus Combined GA

The following graph in Fig. 6 and its corresponding data in Table 5 are presented below. Although, we have generated data of 14 generations for both of these algorithms in this work, this table here displays few feasible solutions and the corresponding computation time for both generic GA and proposed combined GA.

The above graph compares between generic and combined GA for different number of feasible solution. In the above graph of our experiment, it is visible that every

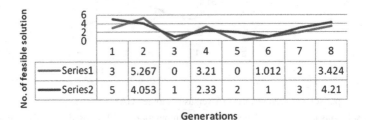

Comparison between Generic GA Vs. Combined GA

	1	2	3	4	5	6	7	8
Series1	3	5.267	0	3.21	0	1.012	2	3.424
Series2	5	4.053	1	2.33	2	1	3	4.21

Generations

Fig. 6 Graphs of Feasible solution and computation time of generic GA versus combined GA

time the number of feasible solutions is obtained in higher value than the combined genetic algorithm.

6 Conclusion

In this research paper, a framework for nurse assignment is described for both home-care and hospital care services. Several research issues are described on basis of which we have proposed a variable time quantum based nurse scheduling (VQTNS) algorithm for both scheduling and assigning of suitable nurses by mapping their grades to home patient's varying conditions. Parameters like dynamic waiting time, swapping time, and critical time are being considered here to develop a cost function in a periodic nurse scheduling. The proposed VQTNS algorithm is compared to other pre-existing algorithms in respect of the computed time and cost. It is observed that our algorithm produces better results than traditional genetic algorithm. In emergency service, the proposed algorithm implements nearest neighbour joining similarity measure to substitute the appropriate nurses in case of her absence with the help of previous week's schedule and produces cost function in reduced time compared to other existing work. A combined genetic algorithm with local search is used to generate the best feasible shift pattern in case of regular normal situation in hospital care service. Finally, the results show that our proposed combined GA produces more number of feasible solutions in less time than traditional GA. Still there is a scope to validate the proposed framework in future and to implement this work in a large-scale application.

References

1. Li, J., Aickelin, U.: Bayesian optimisation algorithm for nurse scheduling, scalable optimization via probabilistic modeling: from algorithms to applications. In: Pelikan, M., Sastry, K., Cantu-Paz, E. (eds.) Studies in Computational Intelligence (Chapter 17), pp. 315–332. Springer, Berlin (2006)
2. Aickelin, U., Downsland, K.A.: Exploiting problem structure in genetic algorithms approach to a nurse rostering problem. J. Sched. **31**, 139–153 (2000)
3. Maenhout, B., Vanhoucke, M.: Comparison and hybridization of crossover operators for the nurse scheduling problem. Ann. Oper. Res. **159**(1), 333–353 (2008). https://doi.org/10.1007/s10479-007-0268-z
4. Fonseca, G.H., Santos, H.G., Carrano, E.G.: Late acceptance hill-climbing for high school timetabling. J. Sched., 1–13 (2015). https://doi.org/10.1007/s10951-015-0458-5
5. Burke, E.K., Newall, J.P., Weare, R.F.: A memetic algorithm for university exam timetabling. genetic algorithms practice and theory of automated timetabling. Lecture Notes in Computer Science, vol. 1153, pp. 241–250. Springer, Berlin (2005)
6. Constantino, A.A., Landa-Silva, D., Melo, E.L., Xavier de Mendonc, D.F., Rizzato, D.B., Romão, W.: A heuristic algorithm based on multi assignment procedures for nurse scheduling. Ann. Oper. Res. **218**(1), 165–183 (2014). https://doi.org/10.1007/s10479-013-1357-9
7. Brucker, P., Burke Edmund K., Curtois, T., Qu, R., Berghe, V.G.: A shift sequence based approach for nurse scheduling and a new benchmark dataset. J. Heuristics **16**(4), 559–573 (2010)
8. Maenhout, B., Vanhoucke, M.: An electromagnetic meta-heuristic for the nurse scheduling problem. J. Heuristics **13**(4), 359–385 (2007)
9. Ratnayaka, R.K.T., Wang, Z.J., Anamalamudi, S. and Cheng, S.: Enhanced greedy optimization algorithm with data warehousing for automated nurse scheduling system. E-Health Telecommun. Syst. Netw. **1**, 43–48 (2012). http://dx.doi.org/10.4236/etsn.2012.14007
10. Elahipanah, M., Desaulniers, G., Lacasse-Guay, È.: A two-phase mathematical-programming heuristic for flexible assignment of activities and tasks to work shifts. J. Sched. **16**(5), 443–460 (2013). https://doi.org/10.1007/s10951-013-0324-2
11. Alkan, A., Ozcan, E.: Memetic algorithms for timetabling. In: The 2003 Congress on Evolutionary Computation, CEC '03, vol. 3, pp. 1796–1802. IEEE (2003). https://doi.org/10.1109/cec.2003.1299890
12. Ko, Y.W., Kim, D.H., Jeong, M., Jeon, W., Uhmn J., Kim, J.: An improvement technique for simulated annealing and its application to nurse scheduling problem. Int. J. Softw. Eng. Appl. **7**(4), (2013)
13. Runarsson, T.P., Yao, X.: Stochastic ranking for constrained evolutionary optimization. IEEE Trans. Evol. Comput. **4**(3), 284–294 (2000)
14. Bai, R., Burke, K.E., Kendall, G., Li, J., McCollum, B.: A hybrid evolutionary approach to the nurse rostering problem, evolutionary computation. IEEE Trans. Evol. Comput. **14**(4), 580–590 (2010). ISSN: 1089-778X
15. Dias, T.M., Ferber, D.F., de Souza, C.C., Moura, A.V.: Constructing nurse schedules at large hospitals. Int. Trans. Oper. Res. **10**, 245–265 (2003)
16. Aickelin, U., Dowsland, K.A.: An indirect genetic algorithm for a nurse-scheduling problem. Comput. Oper. Res. **31**(5), 761–778 (2004)
17. Moz, M., Pato, M.V.: A genetic algorithm approach to a nurse rerostering problem. Comput. Oper. Res. **34**, 667–691 (2007). https://doi.org/10.1016/j.cor.2005.03.019
18. Needleman, J., Buerhaus, P., Mattke, S., Stewart, M., Zelevinsky, K.: Nurse-staffing levels and the quality of care in hospitals. N. Engl. J. Med. **346**, 1715–1722 (2002). https://doi.org/10.1056/nejmsa02247
19. Tsai, C., Li, A.H.S.: A two-stage modeling with genetic algorithms for the nurse scheduling problem. Expert Syst. Appl. **36**, 9506–9512 (2009)
20. Aickelin, U., White, P.: Building better nurse scheduling algorithms. Ann. Oper. Res. **128**(1), 159–177 (2004). https://doi.org/10.1023/b:anor.0000019103.31340.a6

21. Moscato, P., Cotta, C.: A modern introduction to memetic algorithms (Chapter 6). In: Gendreau, M., Potvin, J.-Y. (eds.) Handbook of Metaheuristics, International Series in Operations Research & Management Science, vol. 146, pp. 141–183. Springer, US (2010). https://doi.org/10.1007/978-1-4419-1665-5

22. Aickelin, U.: An indirect genetic algorithm for set covering problems. J. Oper. Res. Soc. **53**(10), 1118–1126 (2002)

23. Saitou, N., Nei, M.: The neighbor-joining method: a new method for reconstructing phylogenetic trees. Mol. Biol. Evol. **4**(4), 406–425 (1987)

A Novel Bio-inspired Algorithm for Increasing Throughput in Wireless Body Area Network (WBAN) by Mitigating Inter-WBAN Interference

Sriyanjana Adhikary, Samiran Chattopadhyay and Sankhayan Choudhury

Abstract Wireless Body Area Network (WBAN) is a type of Wireless Sensor Network (WSN) that has tremendous applications in the healthcare domain. Transmission of the critical health data must be prioritized over periodic ones to deliver it within the stipulated time. Further, as the spectrum is limited and fixed, there is a contention among users for its usage. The objective of this paper is to maximize the system throughput by allocating the available bandwidth fairly based on the criticality of the data and mitigating inter-WBAN interference in a densely populated network by optimizing the transmission power. We have mathematically modeled the problem as a Linear Programming Problem (LPP) and solved it using Particle Swarm Optimization (PSO). Simulation results show that the proposed solution converges quickly. It also outperforms 802.15.6 and other existing state-of-the-art algorithms in terms of throughput, packet delivery ratio and retransmission of packets in high mobility scenarios.

Keywords Wireless Body Area Network · Inter-WBAN interference · PSO

1 Introduction

The advancement in information, communication, embedded system and sensor world, a promising technique called Wireless Body Area Network (WBAN) evolved to challenge the traditional model and revolutionize the health and medical sector by exchanging information between a person and a distantly located control center

S. Adhikary (✉) · S. Chattopadhyay
Department of Information Technology, Jadavpur University, Kolkata, India
e-mail: sriyanjana@gmail.com

S. Chattopadhyay
e-mail: samirancju@gmail.com

S. Choudhury
Department of Computer Science & Engineering, University of Calcutta, Kolkata, India
e-mail: sankhayan@gmail.com

© Springer Nature Singapore Pte Ltd. 2018
R. Chaki et al. (eds.), *Advanced Computing and Systems
for Security*, Advances in Intelligent Systems and Computing 667,
https://doi.org/10.1007/978-981-10-8183-5_2

[1–3]. WBAN comprises of tiny wireless sensors and one or more coordinators that are positioned in, on or around the human body. Body sensors are capable of collecting health parameters and transmit it to the coordinator in a wireless manner. These coordinators convey the health data to some Access Point (AP) wirelessly. Through these APs, data can be propagated to some remote location for further analysis and diagnosis. This system has the prospect to manage wellness of the society rather than curing illness. The potential of WBAN can be utilized and accepted by the society only if health parameters can be continuously monitored, transmitted securely and reliably with minimum delay and without losing critical information despite the presence of dynamic radio propagation. Mobility of a human body aggravates the dynamic nature of radio propagation in WBAN. Further, the radio signals also have adverse effects on different human tissues in the frequency range between 10 kHz and 1 GHz [4]. Mehfuz et al. [4] also explain that the signal quality gets affected by the transmission around human body in accordance with channel model. In [5], requirements of WBASNs based on the IEEE 802.15.6 standard are summarized as follows. The bit rate of a link has to be in the range of 10 kbps–10 Mbps; the packet error rate should be less than 10% for a 256 octet payload for 95% of links, and the time to join or leave a network has to be less than 3 s. The communication architecture of WBASNs is divided into three tiers: intra-WBASN communication (tier 1), inter-WBASN communication (tier 2), and beyond-WBASN communication (tier 3) as shown in Fig. 1.

Human beings actions and movements are unpredictable. Thus, the change in network topology of the network is very high. In this scenario, there is a high probability of losing critical health data of a person. This is extremely undesirable in WBAN as it deals with highly crucial health-related data. Moreover, in worst case, this may even be fatal. Thus, the reliability of the network directly affects the quality

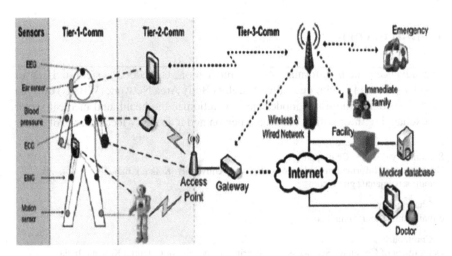

Fig. 1 Three tier architecture of WBAN

of service and the holistic view of the persons health. Reliability ensures quality of service. Reliability can be categorized in two forms: infrastructure and message delivery. Infrastructure reliability indicates continuation of the service even if there is a network failure or a device is unable to access the sink node. On the other hand, reliable message delivery signifies the delivery of the body vitals without colliding with of other signals present in the same frequency range.

In this paper, we focus on message delivery reliability and the quality of service defined in terms of Bit Error Rate (BER) and Packet Delivery Ratio (PDR). We have shown that increased throughput of the network in terms of higher PDR value is achieved. Further, as WBAN sensor nodes are energy constrained, we have ensured the right balance of performance and power consumption by mitigating inter-WBAN interference effectively.

In order to ensure maximum throughput of health data packets, this paper utilizes an emerging promising technology of wireless communication called Overlay Cooperative Cognitive Radio (OC-CR). This technology solves the problem of limited crucial spectrum allocation.

The contributions of this paper are summarized as follows:

1. Allocate the available frequency channels fairly based on the criticality of the health data to be transmitted.
2. Optimize the transmission power in order to achieve higher channel gain by the cognitive users.
3. The proposed algorithm effectively avoids collision and hence enhance the throughput of the system.
4. Number of retransmissions to reframe the network as quick as possible is much less as compared to General 802.15.6.
5. The proposed algorithm outperforms 802.15.6 [6], and [7] in terms of probability of collision and throughput.

The paper is organized as follows: Sect. 2 discusses the current state of the art in this domain followed by an elaborate design model formulation of the proposed work. The problem is solved in Sect. 4 based on Particle Swarm Optimization (PSO). Section 5 demonstrates the simulation results of the proposed algorithm followed by conclusion in Sect. 6.

2 Related Work

Active research on enhancing reliability in WBAN by mitigating inter-interference is still an ongoing process. Interference is categorized in two types: Mutual Interference (MI) and Cross Interference (CI). MI is defined as the existing of radio signal of the same types within the transmission range. CI occurs when radio signals of many technologies having the same frequency are present at the same time. de Silva et al. [8] pointed out via experiments that increasing the number of WBANs adversely affects their PDR. This further degrades by mounting the data rate of the nodes in the network. Though the problem was well defined in the paper, the solution to this problem was not addressed. Wang and Cai [9] analyzed co-channel interference among nonoverlapping WBANs. In this paper, interference is defined in terms of boundary network distance to maintain the signal-to-interference-plus-noise ratio (SINR) threshold. The paper concluded that the minimum network distance should be larger than 7–12.5 m to guarantee the SINR of the boundary nodes. However, this solution restricts human mobility which is not desirable. Different MAC schemes were investigated in [10] with respect to probability of collision, SINR and BER. This paper showed that TDMA outperforms CDMA in co-channel interference mitigation with respect to BER and SINR. But FDMA is known to be the best solution for interference easing in uncoordinated WBANs. Januszkiewicz [11] discussed cognitive radio-based intra-WBAN and inter-WBAN architecture and related issues. It presents an overview of these architectures. Jin et al. [12] used naive Bayesian-supervised learning method to detect and predict coexisting WBANs. Packet Reception Ratio (PRR) and SINR are used for measuring parameters. But the issue of Quality of Service (QoS) is not addressed in this paper.

According to the current state of the art, different schemes to reduce inter-WBAN interference include power control strategies, MAC-based, cognitive radio-based techniques, Ultra-Wideband (UWB) based, and signal processing based approaches. As WBAN sensors are energy constrained, reducing power consumption tones down transmission range and energy usage. As a result, this approach lowers the possible interference area and thus enhances the throughput and lifetime of the network. Proactive Power Update (PAPU) is discussed in [13] that applied game theory to find the optimum transmission power. Similar to [13], Kazemi et al. [14] proposed a game theory-based power control to maximize system throughput while minimizing power expenditure. Here, nodes with high power usage are penalized. A reinforcement learning based on game theory is proposed in [1] for interference mitigation. In this paper, WBAN users transmit at the maximum data rate and the payoff function of one of the users is considered as the reward function to choose the best responding player. Whenever there is a change in the network, this reward function is updated and informed to every node in the network. In [1, 2, 13, 14], huge amount of periodic message exchange takes place in the proposed algorithms to update the status of all nodes in the network. As WBAN nodes are highly mobile and energy constrained, this acts as an unbearable overhead for the lifetime of the network. Kazemi et al. [15] describe a Fuzzy Power Controller (FPC) system model that determines the level of transmission power based on SINR, current power of interference and a feedback channel of the current transmission power. The convergence time of the genetic-fuzzy system is based on the stability of the network. As WBAN is a dynamic system, the number of iterations needed to converge is very high. In [16], an algorithm was designed by detecting network topology and applying PCG. In this mechanism, Bluetooth and acoustic wave technology are used to develop dynamic interaction graph in social network. Implementing this algorithm in a WBAN node is highly impractical because of energy constraint. In addition, the time to detect the distance between two WBANs using acoustic wave is not acceptable in this type of network. Moreover, accurate detection of the wireless channel state is still a research problem.

Zou et al. [17] elaborate another scheme to lessen intrusion by maximizing the payoff of each WBAN that is evaluated by the Bayesian equilibrium. In reality, mapping each WBAN to a set of actions based on its type is a challenging issue. Spanakis et al. [18] describe a Cross-Layer Interference Management (CLIM) scheme in which a WBAN adapts to another transmission rate and power whenever it detects interference. This scheme fails in a crowded network.

In [19], a QoS-based MAC scheduling scheme has been described for healthcare monitoring. This approach considers coexistence of multiple WBANs. The interfering WBANs coordinate among themselves to schedule their transmission time. This is done by exchanging information among them. As WBAN users are highly movable, within the short duration of stability of the network, these extensive information exchanges become unrealistic. Kim et al. [20] use a hybrid approach of CSMA/CA and TDMA in a new MAC algorithm called Asynchronous Internetwork Interference Avoidance (AIIA). They have taken into account a less mobile network. In this method, some sections of a definite WBANs active super-frame [21] are dedicated for coordination. Though it has better results in terms of coordination time, network capacity remains a challenge. Hwang and Kim [22] depict a Continuous Frame Transfer (CFT) of radio module with ultra-low power characteristics. It senses the channel to control the transmission time. Thus, energy is greatly conserved in this method by limiting the number of retransmissions. In Random Incomplete Coloring [7], authors have considered a hard assumption of perfect super-frame synchronization to mitigate inter-WBAN scheduling by graph coloring. In both of [7, 22], delay is not considered at all which is a crucial parameter in healthcare. Two-layer MAC (2L-MAC) protocol [23] uses a polling mechanism to avoid collision while transferring data. If the delay of the responder is higher than a threshold value, it switches to a backup channel. But neither channels wobbliness nor the mobility of WBAN has been addressed. The end-to-end delay is also very high in a medium dense network. Yang and Sayrafian-Pour [6] mitigate interference by adaptive techniques based on Interference Mitigation Factors (IMF). Although the procedure is quite simple, it fails to assign priority of WBANs in different practical situations. Moreover, as in the IMF mechanism, there is no scanning of the channel; thus, collision in the physical layer is highly probable. Cheng and Huang [7] considered all nodes in a WBAN as a whole and all nodes in the same WBAN must work in the same time slot. This assumption is too unrealistic in healthcare domain. This is being overcome in [24] where a Clique-Based WBAN Scheduling (CBWS) is proposed. Each node in a WBAN has the capability of sleep or wake up. This enhances the network lifetime. But this protocol did not consider the channel quality or the priority of any WBAN coordinator based on the type of data. In [6], an Adaptive Internetwork Interference Mitigation takes care of priority of each sensor node for orthogonal scheduling. This ensures the signal QoS at the receivers end. But the end-to-end delay of the signal is considerably high because the sensor nodes can propagate in only one slot and then must pass the time for their next round. Movassaghi et al. [25] describe another MAC approach called Cooperative Scheduling (CS) that has a high spatial reuse because they form clusters to lessen collision. However, priority of sensor nodes and user mobility are not addressed. This approach takes a huge time to converge.

3 WBAN Design Model

3.1 System Model

In this paper, data generated from the body sensors are classified as Throughput Sensitive (TS) and Delay-Sensitive (DS) based on their urgency of delivery. The WBAN coordinators which have DS data are classified as Primary User (PU) by the AP for that particular super-frame transaction and the remaining WBANs participating in the network are Secondary Users (SU). Thus, the tag of PU and SU dynamically changes in the network based on the type of data a WBAN intends to transmit. At a particular time frame, channel gain is maximized by minimizing hindrances, maintaining coordination and collaboration among the network components. To deliver critical data with the best effort to the intended AP, the entire bandwidth available is distributed among the PUs only. The spectrum utilization of the PUs is not constrained so that the earliest delivery of critical data is possible. Here, a non-negotiation scheme is followed to optimize performance and quick convergence. However, each SU is aware of the possible collision with other users in the network by exchanging information among them. This negotiation method enriches our proposed scheme to support dynamic environment. The average human torso is less than 3 m by 3 m. The longest distance a user has to cover to reach its AP is the diagonal of a $\sqrt{3}$ m * 3 m square, which is 3 2 m. To cover this distance, if an AP is not within the transmission range of the PU, it tends to increase its power; this is extremely undesirable in WBAN as it will not only dissipate energy faster, but also cause higher interference. To diminish the effect of interference, the SUs intend to utilize the gaps in transmission in the spectrum dedicated to a particular PU. Since multiple SUs compete to use the underutilized spectrum, possibility of collision exists. To solve this problem, AP makes a pseudo 2-tier architecture. AP acquires appropriate information about the transmission power of the SUs and makes Actual-Level (AL) groups with the conflicting SUs. Any member of a group cannot be allocated in the same spectrum for that particular time frame. Thus, they form different Virtual-Level (VL) groups on the basis of the spectrum in which they are being allowed to transmit. In order to lessen the transmission power of the PU, these SUs forming VL groups may also act as relay nodes.

In this paper, number of nodes in the network is not restricted, provided the allocation of available frequency bandwidth meets the demand of each of the PUs. This is justified because WBAN is a highly mobile network. So the probability of change in density of the network is very high. As a result, the number of PUs also varies at a particular time instant. Further, SUs form VL groups based on the spectrum in which they transmit. This negotiation scheme enriches our proposed algorithm to support scalable dynamic environment.

3.2 Optimization Problem Formulation

The notation used in the formulation is presented in Table 1. The assumptions used in this paper are as follows:

1. $\Sigma B_i(PU)$ less than or equal to the logical bandwidth of the wireless channel.
2. ΣX_{ij} less than or equal to 1; 1 less than or equal to i less than or equal to P and 1 less than or equal to j less than or equal to S;
   This indicates that at the most one SU is allowed to cognitively utilize a channel of a PU at time t.
3. The WBAN is under Additive White Gaussian Noise (AWGN) network using BPSK amplitude modulation to maintain a tradeoff between the delay and the throughput of the received health data.

Maximum Transmit Rate of the jth SU on ith PU channel at time t is given by Eq. (1).

$$\sqrt{r_{i,j}(t)} = B_i \log_2\left(1 + (T_{i,j}(t) \cdot G_{i,j}(t)/Q(2Y_b))\right), \tag{1}$$

where $Q(X) = (1/2\pi) \int_x^\infty e2^{-x}$.

Table 1 Notation

Total available bandwidth	B
Number of PUs at time 't'	P
Number of SUs at time 't'	S
Bandwidth allocated to ith PU at time 't'	$B_i(PU) = B/P$
Available resource information at time 't' $A_{i,j}(t)$	1/0
	=1, when jth SU cannot use ith PU's channel at time 't' because it has already been utilized by ith P =0 otherwise
Channel assignment at time 't' $X_{i,j}(t)$	1/0
	=1, when ith PU channel is assigned to jth SU at time 't' =0 otherwise
Transmission power of jth SU on ith PU channel at time 't'	$T_{i,j}(t)$
Maximum transmission power of sth SU	$T \times S_{max}$
Channel gain of jth SU if it is allocated to use ith PU channel at time 't'	$G_{i,j}(t)$
Maximum transmit rate of jth SU on ith PU channel at time 't'	$r_{i,j}(t)$
Energy transmitted per bit/Noise (SINR)	Y_b

Thus, the optimization problem to maximize throughput of an SU can be written as follows.

$$\text{Max} \sum_{j=1}^{S} \sum_{i=1}^{P} X_{ij}(t) \cdot r_{ij}(t)$$

To complete the mathematical model, the following constraints must be satisfied.

$$\sum_{i=1}^{P} r_{i,j}(t) \geq r_0 \qquad (2)$$

$$\sum_{i=1}^{P} T_{i,j}(t) \leq Tx_{s\text{max}} = 1 \qquad (3)$$

$$0 \leq T_{i,j}(t) \forall 1 \leq i \leq P \quad \text{and} \quad \forall 1 \leq j \leq S \qquad (4)$$

Equation 2 illustrates that the transmission rate of the cognitive users must not go below the threshold value r_0. This ensures fairness in the spectrum allocation. Equation 3 refers to the transmission power constraints. It guarantees that the total transmission powers of all the SUs allocated in PU's channel must be less than or equivalent to the maximum transmission power of the SU's. This makes sure that the transmission power of the SUs is bounded. Equation 4 assures that the transmission power of a SU allocated to a PU's channel must not be negative. The channel allocation takes a discrete value, while the power constraint is of continuous nature. This represents a Mixed Integer Nonlinear Problem (MINP). Because of the complexity in solving this type of problems, this has been transformed into a Continuous Nonlinear Programming Problem (NLP).

4 Problem Solving Based on Particle Swarm Optimization (PSO) Technique

The optimization problem formulated is an NP problem. So the problem has been modeled by a biological evaluation process called Particle Swarm Optimization (PSO) [25–27].

PSO is conceptually simple and is easy to implement. Its control parameters are robust and computationally, it is much more efficient compared to other optimization techniques such as Genetic Algorithm (GA). In GA, information are being shared among chromosomes. So the whole population move like a single group toward an optimal area. PSO uses a unidirectional information sharing mechanism because only global best solution gives out the information to others. In every evolution, it tries to converge toward the best solution. Hence, PSO tends to converge to the best solution quickly.

In PSO, each solution is denoted by a particle that has a position vector and a velocity vector. According to the aforementioned problem, a particle signifies a possible power and bandwidth allocation of the conflicting SUs at time 't'. The allocation focuses on interference mitigation by fair assignment of spectrum to avoid collision and thus enhances throughput of the system.

Moreover, PSO simulates the behavior of flocking of bird. In our problem, incompatible SUs randomly explore frequency channels to propagate TS data to the intended receiver at time 't'. None of the SUs have any idea of the best channel. They only have the knowledge of the distance to the best solution in each iteration. All particles (SUs) determine their searching speed and direction within the search space based on their respective velocities. The performance of each particle depends on its fitness function which is determined by the objective function. The position of the particles at the convergence point gives the solution to the problem.

4.1 Notation

Notation used in this paper for PSO technique are as follows:

- Q is the Number of particles.
- Position of particle $i(1 \leq i \leq Q)$ at the tth iteration is denoted by $P_i(t) = [P_{i1}(t), P_{i2}(t), P_{i3}(t), \ldots P_{iD}(t)]$, where D is the number of dimensions to represent the particle $d \in D - P_i(t)$ is a probable solution.
- Velocity of particle $i(1 \leq i \leq Q)$ at the tth iteration is denoted by $V_i(t) \in [V_{i1}(t), V_{i2}(t), V_{i3}(t), \ldots V_{iD}(t)]$, where $V_{-i}D(t) \in [-V_{max}, V_{max}]$.
- Local Best Solution of the particle i until the tth iteration is $P_{bi}(t) = [P_{bi1}(t), P_{bi2}(t), P_{bi3}(t), \ldots P_{biD}(t)]$.
- Global Best Solution of the particle i until the tth iteration is $P_g(t) = [P_{g1}(t), P_{g2}(t), P_{g3}(t), \ldots P_{gD}(t)]$.
  The value of every bit in the particle is randomly generated in the initial phase. This scheme reduces the search space efficiently and thus the convergence time gets reduced.
- In every iteration, the velocity is updated as follows $V_{id}(t + 1) = wV_{id}(t) + c_1r_1[P_{biD}(t) - P_{iD}(t)] + c_2r_2[P_{gD}(t) - P_{iD}(t)]$, where c_1 and c_2 are constants lying between 0 and 1. These are used to control the amount of variation in velocity. r_1 and r_2 are used to add randomness in the algorithm to explore the algorithm for varying degree. w is used to vary the priority of previous values and the current value. Too high a jump may cross the global maxima while too less a jump may take a long time to converge. Thus, we focused on balancing between exploration and exploitation in the evolutionary computing technique.
- In every iteration, the position is updated as follows.

$$P_{id}(t + 1) = P_{id}(t) + V_{id}(t + 1)$$

- The fitness function is calculated as follows.
  $F_{i,j}(t) = \sum_{j=1}^{S} \sum_{i=1}^{P} X_{ij}(t) \cdot r_{ij}(t)$—the number of times the constraints has been violated.

4.2 PSO Based Spectrum Allocation Algorithm

Step 1: Propagate resource allocation matrix L, channel information matrix $G_{i,j}$, dimensions of each particle, and the total number of population to SUs and subsequent APs.

Step 2: Initialize the number of iterations $t = 0$ and randomly generate $V_i(t)$ and $P_i(t)$ where $V_{iD}(t) \in \{-V_{max}, V_{max}\}$ and thus acquires $P_i(t) = \{P_{i1}(t), P_{i2}(t), P_{i3}(t), \ldots P_{iD}(t)\}$ for each $i(1 \leq i \leq Q)$

Step 3: Associate each $P_{id}(t)$ $(1 \leq i \leq Q)$ to $T_{i,j}(t)$ where (i, j) is the dth element with $A_{i,j}(t) = 1$

Step 4: Compute Fitness Function (CFF) of each particle in the population using the equation $F_{i,j}(t) \sum_{j=1}^{S} \sum_{i=1}^{P} X_{ij}(t) \cdot r_{ij}(t)$—the number of times the constraints has been violated Find Local Best Solution
$P_{bi}(t) = \lfloor P_{bi1}(t), P_{bi2}(t), P_{bi3}(t), \ldots P_{biD}(t) \rfloor$ and Global Best Solution
$P_g(t) = \lfloor P_{g1}(t), P_{g2}(t), P_{g3}(t), \ldots P_{gD}(t) \rfloor$
where b indicates the particle with highest fitness value.

Step 5: Set $t = t + 1$ and Update Velocity using $V_{id}(t + 1) = wV_{id}(t) + c_1 r_1 [P_{biD}(t) - P_{iD}(t)] + c_2 r_2 [P_{gD}(t) - P_{iD}(t)]$
If the current value $V_{id}(t) \geq V_{max}$, set $V_{id}(t) = V_{max}$
If the current value $V_{id}(t) \leq -V_{max}$, set $V_{id}(t) = -V_{max}$

Step 6: Update position vector using
$P_{id}(t + 1) = P_{id}(t) + V_{id}(t + 1)$

Step 7: Computer Fitness Value of each particle in the population
For each particle i, if Fitness Value $\geq$ Local Best Value, set the current fitness value as the Local Best Solution, i.e., $F_{i,j}(t) = [P_{bi1}(t), P_{bi2}(t), P_{bi3}(t), \ldots P_{biD}(t)]$
For each particle, if Fitness Value $\geq$ Global Best Value, set the current fitness value as the Global Best Solution, i.e., $F_{i,j}(t) = [P_{g1}(t), P_{g2}(t), P_{g3}(t), \ldots P_{gD}(t)]$

Step 8: If $t = $ max iteration value, then stop and output the Global Best Value

$$[P_{g1}(t), P_{g2}(t), P_{g3}(t), \ldots P_{gD}(t)] \text{ to } P_i(t)$$

Else go to Step 5

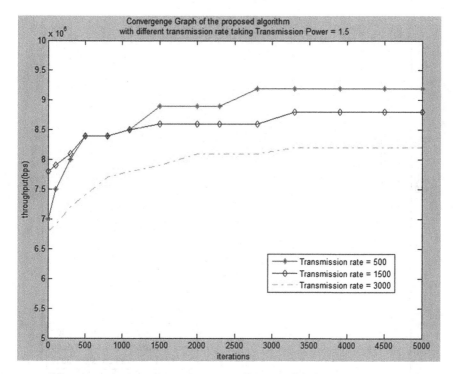

Fig. 2 Convergence graph for transmission power 1.5

5 Simulation Results

To assess the proposed algorithm, it is simulated in MATLAB. It is evaluated based on a cognitive radio system having 12 primary users and 10 secondary users using 6 MHz available bandwidth. The matrix of resource information is generated randomly. For a wireless link, the average channel gain of a cognitive user is randomly generated between 0 and 0.02. The interfering noise is 0.6 mW. The total number of particles in the population is considered to be 20. To conflow the amount of variation velocity possible, the variables $c1$ and $c2$ are taken as 2. The maximum velocity is assumed to be 5. The different transmission rates considered are 500, 1500, and 3000 bps. To compare the transmission power, we have taken into account two cases with 1.5 and 2.5 mW. Both Figs. 2 and 3, demonstrate that the proposed algorithm gives optimum results after 2500 iterations. From the figures, we can also conclude that the throughput of the system does not vary only with the transmission rate. It is constrained by the power with which users are transmitting. Thus, cognitive users having higher channel gains are seen to be using more channels dissipating more power than users having less channel gain but more available power. Hence, the latter category can capture the remaining channels to transmit data. As a result, overall performance of the system improves increasing the throughput of the system.

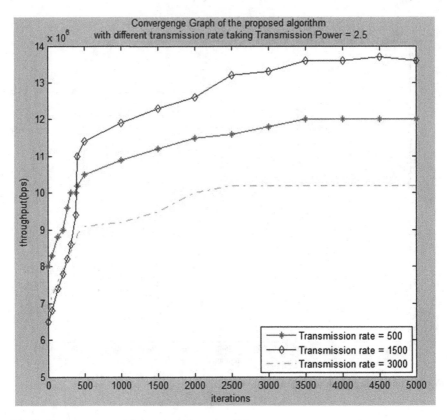

Fig. 3 Convergence graph for transmission power 2.5

Figure 4 shows graph of throughput versus the number of nodes in General 802.15.6. It shows that as the number of nodes in the network increases, the probability of failure enhances than the success rate. Figure 5 shows that using our Overlay-CR algorithm, the transmission success rate is almost constant irrespective of presence of interfering nodes. From the last two graphs, we can conclude that the number of packets delivered to the receiver using our algorithm is at least 20% more than that of the general 802.15.6. This shows that our algorithm outperforms the general one to mitigate interference and hence increases reliable packet delivery.

The overhead of the proposed algorithm is also depicted in Fig. 6 based on the number of retransmission of packets. From the graph, it is clear that the number of retransmissions of packets in the proposed algorithm is comparatively much less than 802.15.6. This is also predictable since the SUs that have high remaining energy act as relay nodes; thus, the network becomes stable very fast even in the presence of high mobility.

Figure 7 compares the proposed algorithm with [6]. The probability of collision in [6] is much higher than the proposed one because the [6] considered adaptive schemes

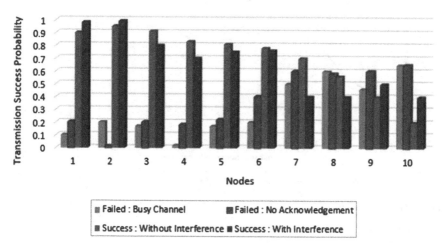

Fig. 4 Throughput of 802.15.6

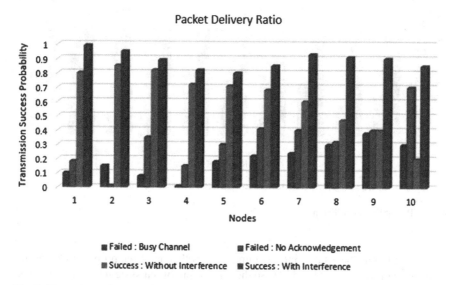

Fig. 5 Throughput of proposed algorithm

that perform better only in low populated channels. When the number of competitive users increases, the performance of their algorithm degrades. The proposed algorithm performs better in real-life scenarios where the population varies dynamically.

Figure 8 demonstrates the system throughput of the proposed algorithm compared to [7]. As Cheng and Huang [7] maintained a perfect super-frame synchronization, there is no scope to support dynamic change in the environment. Thus, priority of

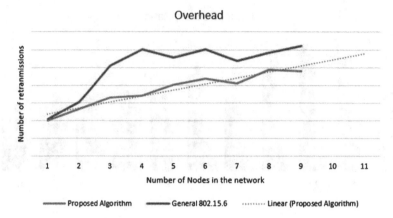

Fig. 6 Overhead comparison graph

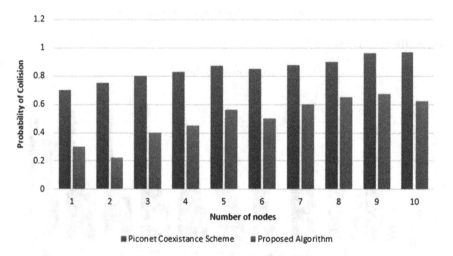

Fig. 7 Probability of collision

health data has not been taken into account. On the other hand, our proposed algorithm prioritized the users as SU and PU. So the throughput of our system is better in terms of PDR.

Table 2 illustrates comparison of the proposed algorithm with different state-of-the-art algorithms based on different parameters.

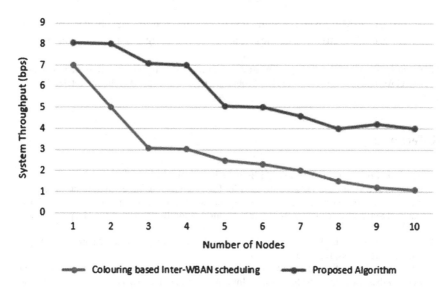

Fig. 8 System throughput probability compared to coloring based inter-WBAN scheduling

Table 2 Comparison table

Parameters versus Algorithms	Proposed Algorithm	General 802.15.6	Piconet coexistence scheme	Coloring based inter-WBAN scheduling
Overhead	Moderate	High	High	High
Probability of Collision	Low	High	Moderate	Moderate
System throughput (bps)	High	Low	Moderate	Moderate

6 Conclusion

In this paper, we have proposed a mathematical model-based algorithm to mitigate inter-WBAN interference. In our algorithm, we have applied the concept of Overlay-CR system. As WBAN deals with highly dense and frequently changing network, we have considered a AWGN network. To verify and validate the model, we have proposed an algorithm using PSO. Performance of the algorithm has been measured through simulations in MATLAB. Simulation has been done varying the transmission rate and power. Simulation results suggest that with 1.5 mW transmission power, when the transmission rate is nearly 3000 bps, the throughput of the system slows down. This result is expected because higher transmission rate ends up in data loss. But when the transmission power is increased to 2.5 mW and the transmission rate as 3000 bps, the throughput of the system increases by at least 20% as compared to the general 802.15.6. Moreover, the number of retransmissions of packets in the proposed algorithm is comparatively much less than 802.15.6 in high mobility scenarios.

Acknowledgements This work is supported by Information Technology Research Academy (ITRA), Government of India under, ITRA-Mobile grant [ITRA/15 (59)/Mobile/RemoteHealth/01].

References

1. Aziz, O., Lo, B., Darzi, A., Yang, G.: Body Sensor Network (2011)
2. Kazemi, R., Vesilo, R., Dutkiewicz, E., Liu, R.P.: Reinforcement learning in power control games for internetwork interference mitigation in wireless body area networks. In: 2012 International Symposium on Communications and Information Technologies (ISCIT), pp. 256–262. IEEE (2012)
3. Yuce, M.R., Khan, J.: Wireless Body Area Networks: Technology, Implementation, and Applications. CRC Press, Boca Raton (2011)
4. Mehfuz, S., Urooj, S., Sinha, S.: Wireless body area networks: a review with intelligent sensor network-based emerging technology. In: Information Systems Design and Intelligent Applications, pp. 813–821. Springer, Berlin (2015)
5. Movassaghi, S., Abolhasan, M., Lipman, J., Smith, D., Jamalipour, A.: Wireless body area networks: a survey. IEEE Commun. Surv. Tutor. **16**(3), 1658–1686 (2014)
6. Yang, W.-B., Sayrafian-Pour, K.: Interference mitigation for body area networks. In: 2011 IEEE 22nd International Symposium on Personal, Indoor and Mobile Radio Communications, pp. 2193–2197. IEEE (2011)
7. Cheng, S.H., & Huang, C.Y.: Coloring-based inter-WBAN scheduling for mobile wireless body area networks. IEEE Trans. Parallel Distrib. Syst. **24**(2), 250–259 (2013)
8. de Silva, B., Natarajan, A., Motani, M.: Inter-user interference in body sensor networks: preliminary investigation and an infrastructure-based solution. In: 2009 Sixth International Workshop on Wearable and Implantable Body Sensor Networks, pp. 35–40. IEEE (2009)
9. Wang, X., & Cai, L.: Interference analysis of co-existing wireless body area networks. In: Global Telecommunications Conference (GLOBECOM 2011), IEEE, pp. 1–5. IEEE (2011)
10. Zhang, A., Smith, D.B., Miniutti, D., Hanlen, L.W., Rodda, D., Gilbert, B.: Performance of piconet co-existence schemes in wireless body area networks. In: 2010 IEEE Wireless Communication and Networking Conference, pp. 1–6. IEEE (2010)
11. Januszkiewicz, L.: Simplified human body models for interference analysis in the cognitive radio for medical body area networks. In: 2014 8th International Symposium on Medical Information and Communication Technology (ISMICT), pp. 1–5. IEEE (2014)
12. Jin, Z., Han, Y., Cho, J., Lee, B.: A prediction algorithm for coexistence problem in multiple-WBAN environment. Int. J. Distrib. Sens. Netw. **2015**, 7 (2015)
13. Fang, G., Dutkiewicz, E., Yu, K., Vesilo, R., Yu, Y.: Distributed inter-network interference coordination for wireless body area networks. In: Global Telecommunications Conference (GLOBECOM 2010), 2010 IEEE, pp. 1–5. IEEE (2010)
14. Kazemi, R., Vesilo, R., Dutkiewicz, E., Fang, G.: Inter-network interference mitigation in wireless body area networks using power control games. In: 2010 International Symposium on Communications and Information Technologies (ISCIT), pp. 81–86. IEEE (2010)
15. Kazemi, R., Vesilo, R., Dutkiewicz, E.: A novel genetic-fuzzy power controller with feedback for interference mitigation in wireless body area networks. In: IEEE 73rd Vehicular Technology Conference (VTC Spring), 2011, pp. 1–5. IEEE (2011)
16. Zhang, Z., Wang, H., Wang, C., Fang, H.: Interference mitigation for cyber-physical wireless body area network system using social networks. IEEE Trans. Emerg. Top. Comput. **1**(1), 121–132 (2013)

17. Zou, L., Liu, B., Chen, C., Chen, C.W.: Bayesian game based power control scheme for inter-WBAN interference mitigation. In: 2014 IEEE Global Communications Conference, pp. 240–245. IEEE (2014)
18. Spanakis, E.G., Sakkalis, V., Marias, K., Traganitis, A.: Cross layer interference management in wireless biomedical networks. Entropy **16**(4), 2085–2104 (2014)
19. Jamthe, A., Mishra, A., Agrawal, D.P.: Scheduling schemes for interference suppression in healthcare sensor networks. In: 2014 IEEE International Conference on Communications (ICC), pp. 391–396. IEEE (2014)
20. Kim, E.-J., Youm, S., Shon, T., Kang, C.-H.: Asynchronous inter-network interference avoidance for wireless body area networks. J. Supercomput. **65**(2), 562–579 (2013)
21. Fukuya, T., Kohno, R.: QoS-aware super frame management scheme based on ieee 802.15. 6. In: 2016 10th International Symposium on Medical Information and Communication Technology (ISMICT), pp. 1–4. IEEE (2016)
22. Hwang, H., Kim, N.M.: An enhanced frame transmission method for health devices with ultra low power operation. In: 2012 IEEE International Conference on Consumer Electronics (ICCE) (2012)
23. Chen, G.-T., Chen, W.-T., Shen, S.-H.: 2L-MAC: A MAC protocol with two-layer interference mitigation in wireless body area networks for medical applications. In: 2014 IEEE International Conference on Communications (ICC), pp. 3523–3528. IEEE (2014)
24. Xie, Z., Huang, G., He, J., Zhang, Y.: A clique-based WBAN scheduling for mobile wireless body area networks. Procedia Comput. Sci. **31**, 1092–1101 (2014)
25. Movassaghi, S., Abolhasan, M., Smith, D., Jamalipour, A.: Aim: adaptive internetwork interference mitigation amongst co-existing wireless body area networks. In: 2014 IEEE Global Communications Conference, pp. 2460–2465. IEEE (2014)
26. Kennedy, J.: Particle swarm optimization. In: Encyclopedia of Machine Learning, pp. 760–766. Springer, Berlin (2011)
27. Shi, Y., Eberhart, R.: A modified particle swarm optimizer. In: The 1998 IEEE International Conference on Evolutionary Computation Proceedings, 1998. IEEE World Congress on Computational Intelligence, pp. 69–73. IEEE (1998)

Part II
Cloud Computing

Paving the Way for Autonomic Clouds: State-of-the-Art and Future Directions

Christina Terese Joseph and K. Chandrasekaran

Abstract Cloud Computing is the core technology that helps in catering to the computing needs of the current generation. As the customers increase, data center providers are looking for efficient mechanisms to handle the management of the large reservoir of resources involved in the Cloud environment. In order to support efficient managing, it is the need of the day to adopt the concept of Autonomic Computing into Cloud. Several researchers have been attempted to improve the managing capability of the Cloud, by encorporating autonomic capabilities for resources in the Cloud. Most of the researches attempt to automate some aspects while the remaining portion of the Cloud does not have autonomic functionalities. An autonomic Cloud is one where all the operations can be handled without human intervention. There is a long way to go to achieve this vision. In our study, we first categorize the various existing approaches on the basis of the methodology employed and analyze the different self-*properties considered by the different approaches. It is observed that in each approach, researchers focus on one or at most two self-*properties. Based on our analysis, we suggest some of the future directions that can be paved on by researchers working in this domain.

1 Introduction

In today's digital era, technological advancements are being made at a rapid pace. Most of the current world solutions to meet the computing needs of the world involve the use of Cloud computing. Cloud computing with its various characteristics such as the pay-as-you-go model is best suited for the increasing computing demands of people. As more and more customers tend to use Cloud, the challenges faced by the

C. T. Joseph (✉) · K. Chandrasekaran
Department of Computer Science and Engineering, National Institute
of Technology Karnataka, Surathkal 575025, India
e-mail: xtina1232@gmail.com

K. Chandrasekaran
e-mail: kchnitk@gmail.com

© Springer Nature Singapore Pte Ltd. 2018
R. Chaki et al. (eds.), *Advanced Computing and Systems
for Security*, Advances in Intelligent Systems and Computing 667,
https://doi.org/10.1007/978-981-10-8183-5_3

Cloud management team are increasing. It is evident that management relying solely on human effort will not be sufficient to cope up with the advancements being made. In order to relieve the human managers, it is necessary to encorporate the Cloud resources with capabilities to govern its own actions. Such a Cloud environment where the resources can manage themselves is called Autonomic Cloud Computing.

The Autonomic Computing Concept was developed by Paul Horn in 2001. The idea basically involved the building of computing systems that can manage themselves, by adapting to the variations in its environment. According to IBM, autonomic systems can be characterized by eight properties, referred to as the self-*properties [12]. The core idea of these eight characteristics can be encapsulated by four self-*properties. To develop self-managing systems, the systems should be self-optimizing, self-configuring, self-healing, and self-protecting. It may be observed that these properties are not completely orthogonal. For example, self-optimizing systems are sometimes self-configuring as well. Apart from these, the literature has proposed many other self-*properties. Berns et al. [1] discuss the scope of each of these properties and attempt to give a formal definition for each of the relevant self-*property. To realize the self-*properties, different actions of the autonomic control loop will have to be carried out. They are Monitor (M), Analyze (A), Plan (P), and Execute (E) [9]. The autonomic control is thus sometimes referred to as the MAPE loop.

Though autonomic computing systems is the holistic vision of automating the entire system, researchers usually decompose the system into various blocks and attempt to encorporate managing capabilities to the individual components. This is not sufficient in building autonomic systems, but is a necessary step.

In this study, we attempt to categorize the relevant contributions to the field of Autonomic Cloud Computing. While performing the study, we have:

- Considered works that implement all of the actions in the autonomic control loop; works that concentrate only on one action of the autonomic control loop are not within the scope of our study.
- Considered research contributions that actually encorporate autonomic capabilities to the system, rather than considering all systems that claim to be autonomic.
- Included some works that provide descriptions of the processes that can be used to support the autonomic management.

The paper is organized as follows: in Sect. 2, we present the approaches considered in this study. Section 3 details on the findings of our study. On the basis of these findings, we suggest some future directions in Sect. 4. The concluding remarks are included in Sect. 5.

2 Review of Existing Approaches in the Literature

Our study considers few relevant contributions to the autonomic resource management in Cloud. We review the approaches based on their objectives, methodology used, the novel contribution of their approach and the validation technique employed. Based on the methodology used to provide autonomicity, we have categorized the approaches into metaheuristic-based approaches, machine learning based approaches, market-based approaches, cryptography-based approaches, hybrid approaches, and the other approaches. The subsections that follow detail on the studies that fall under each category.

2.1 *Metaheuristic-Based Approaches*

This subsection describes the approaches that employ different metaheuristic techniques to improve the managing capability of various resources in Cloud.

User demands are dynamic in Cloud computing. It is the responsibility of the Cloud provider to be able to meet even its peak demands, while at the same time ensuring a decent level of QoS. Zuo et al. [30] propose an approach to guarantee that a provider of IaaS services can meet user demand even under peak loads, without adversely affecting the QoS. The authors propose that the provider can take the assistance of external clouds in scenarios where the capacity of the provider is not adequate to meet the users' needs. The proposed approach eliminates the need for formal agreements that are required in the case of federated Cloud. The optimization problem is formulated as an Integer Programming model and solved using Particle Swarm Optimization. The component responsible for scheduling allocates each task either within the Cloud or to one of the external cloud providers, depending on the availability of resources with the Cloud provider. To choose the most suitable external cloud, their pricing strategy is considered.

Service-Oriented Computing requires for different services to be composed. The composed set of services will have to satisfy the specified level of QoS. The composition will exhibit a level of QoS that is the aggregate of the QoS of the constituents of the composition. Thus, in order to modify the QoS, it is sufficient to change one of the constituent services. Developing an optimal composition of services with the required level of QoS is a challenging issue. Klein et al. [17] observed that the composition generally exhibits a different level of QoS than the aggregated value of the QoS of the constituent services. This is due to the network latency, which is not considered in the approaches proposed so far. The authors thus propose an approach that considers this and chooses the services with the least network latency. For this, they build a network model to estimate the latency of the network between the users and the services. The authors propose an approach that makes use of Genetic Algorithm to calculate an optimum between the latency and the other QoS. The approach adapts itself by dynamically determining the probabilities of the various operators.

As the number of clients opting for Cloud computing increases, the number of malicious clients in Cloud computing is also increasing. A particular case of the security issue in the Cloud environment deals with coresidence profiling. In this type of attack, the VM of a particular customer is targeted. Once the malicious user succeeds in starting his VMs on the same host as that of the target customer, it requires less efforts to obtain sensitive information from the target VM. This class of attacks can be prevented to an extent by adopting some means to ensure that the VMs of the malicious user and the target user is never deployed on the same physical machine. Rice et al. [23] have developed an approach that uses this concept. The approach proposed by the authors is inspired from the behavior of the mussels. In order to be able to withstand the water pressure, the mussels form clusters and stay together. In a similar manner, the users are grouped on the basis of their preferences and requirements, so that they may be placed together thus preventing any malicious users VMs being placed on the same host. Another precaution that the approach takes against the attack is that for each cluster, a proxy address is used. The mapping to the private IP addresses is thus kept hidden from the outside world. In addition to this, the clusters are periodically dissolved and recomputed.

2.2 Machine Learning Based Approaches

Nowadays, many systems are designed such that they have the ability to learn by themselves. Initially such systems required a pre-built model or some sample scenarios in the training phase. The advancements in the machine learning field have resulted in approaches that require no such training. The systems themselves apply different inputs, analyze the error involved, and tune themselves accordingly to produce the desired output. There are different approaches that adopt machine learning techniques to provide entities in Cloud with autonomic capabilities. Such approaches are discussed in this subsection.

With the gaining popularity of Cloud Computing, it is crucial for Cloud providers to ensure QoS level required by the clients. Owing to the dynamic nature of Clouds, the QoS of services provided by the Cloud provider also varies. With time, not only the QoS level, but even the factors affecting the QoS might change. Chen et al. [6] propose a system that is capable of relating the QoS value to the different attributes of the Cloud. They dynamically determine the factors that are determining the QoS level and attempt to tune these factors so that the required QoS is attained. Thus, the proposed model comprises two phases: the primitives selection phase, where the parameters affecting the QoS is determined and the QoS function construction phase, where the relation between the determined parameters and the QoS level is expressed as a mathematical function. A significant contribution of the work is the use of hybrid dual learners. The authors observed that the same learning algorithm cannot be used in all cases of QoS function construction. As the parameters change, the learning algorithm used also changes. Thus, analysis is carried out to study which learning algorithm is more suited to model the relation between the QoS level and the different parameters. The machine learning algorithms considered are ANN, Regression Tree, and ARMAX. The error between models is measured using the SMAPE metric. In order to measure the importance that should be given in a QoS function, the Symmetric Uncertainty concept from information theory has been used.

Autonomic managers are responsible for the implementing various actions of the MAPE control loop. In order to perform these actions efficiently, the managers require some knowledge about the context in which they are deployed. The work by Uriarte et al. [25] tries to provide some of this knowledge to the autonomic managers. They try to differentiate among the different types of services. The random forest (RF) algorithm is used to calculate how similar are the different services. During the training phase, the RF algorithm employs bootstrap aggregating. The result of this algorithm is then passed to the PAM clustering algorithm. Finally we get different clusters, each containing similar services. This knowledge can then be utilized for scheduling tasks, to ensure that similar services are run on different nodes. The major strengths of the proposed approach are that it can consider larger number of features and that it can support online operation.

Autoscaling is one of the mechanisms by which Cloud providers are able to meet its customers demand. Thus, it is necessary that autoscaling be performed only after taking certain factors into consideration. When different services are running on the Cloud, there may be interferences which can affect the QoS level. Consequently, the QoS models that are static cannot be relied upon. There is a need for analyzing the environment and determining how various factors affect the QoS. Chen et al. [7] suggest that there is no machine learning algorithm that is apt for all situations. Instead, the learning algorithm will also have to be decided by considering the scenario. The capability for selecting the most suited learning algorithm can be attained using meta-self-awareness. This feature can then be extended to Cloud federations where it is more challenging to determine what factors are affecting the QoS and how, and then use this knowledge to control the autoscaling performed.

2.3 Market-Based Approaches

Some researchers have considered the Cloud environment to be similar to the market environment, where there are sellers and potential buyers. The sellers and buyers themselves come into an agreement that is most suitable to their requirements. In this subsection, the works that adopt market-based techniques, such as auctions, are discussed.

Applications based on services generally require the selection of the different constituent services. Since the Cloud is a dynamic environment, there may be cases where the constituents will have to be reselected. Nallur and Bahsoon [22] have proposed an adaptive market-based approach for the reselection purpose. The concept of continuous double auction has been utilized here. Both the service providers and the customers send bids. Thus for each service, there will multiple auctions carried out. The bid generation process is guided by the nature of the QoS attributes. The service will be provided to the potential buyer that has agreed to pay a price higher than the price that it has bid, provided all the QoS conditions are satisfied. Thus, we have a distributed process for decision-making.

Cloud Computing decouples the execution environment from the clients locations. This sometimes leads to a deterioration in the quality of service as experienced by the users. Thus, recent developments in the Cloud computing environment include an attempt to bring execution closer to the users. A challenging issue in this scenario is that the executing units will have lesser amount of resources. Thus, tasks such as resource allocation should be done, taking this into consideration. Landa et al. [18] propose an approach to carry out this task. In this approach, the units available for executing are partitioned into execution zones, with each zone being controlled by a zone manager. The applications to be run are managed by application managers. The resource allocation problem can thus be represented as the allocation of execution zones to various application managers. According to the corresponding policies, each application manager tries to associate with a suitable execution zone. For this, a market-based Vickrey auction method is used. The application manager sends its bids for the corresponding zones and the decision is taken by the zone manager who makes the decision by running a Vickrey auction. The system is also provided with the ability to cope up with the changes in the environment by tuning itself.

2.4 Cryptography-Based Approaches

It is generally required that Clouds are secure. The self-*property that provides this is self-protecting. The self-protecting techniques generally employ techniques based on cryptography. This subsection deals some such approaches.

As the popularity of Cloud Computing is increasing, the Cloud provider is endowed with more responsibilities. There is an increase in the number of clients who upload their personal data to the Cloud. It becomes a challenging task for the Cloud provider to protect the users' private data from malicious users. Zeng et al. [28] propose an approach that automatically destroys the data after a time specified by the user. Cryptographic techniques coupled with active storage techniques have been used in order to ensure that the data cannot be retrieved after a specified time. All this is done without the intervention of the user. An active storage framework associates each element with a time period during which it can be accessed. Thus, this approach aims at protecting the system from intruders.

Nowadays, a lot of sensitive data is being shared on the servers in Cloud computing. To ensure that such data is not compromised, Xiong et al. [27] propose a Key-Policy Attribute-based encryption with time-specified attributes (KP-TSABE). Here, instead of associating the stored data with a time interval, each ciphertext has a corresponding time interval while each private key has a corresponding time instant. Decryption of the ciphertext is possible only if the time instant falls within the time interval. The authors attempt to provide a fine-grained control of access. The proposed scheme has been proved to be secure using the l-bilinear Diffe–Hellman Inversion. The authors observe that even if the stored data is deleted from the servers, it is easy to recover it from the physical storage of the Cloud. Hence, they propose such an approach which does not rely on the deletion of data. The time period can be specified by the user. The data cannot be retrieved at any instant outside the specified interval (neither before nor after).

2.5 Hybrid Approaches

Some approaches couple two or more methodologies to improve the performance of the system. Our study classifies these approaches as hybrid approaches. In this subsection, some hybrid approaches to Autonomic Cloud Computing are discussed.

In a Virtualized Cloud Data Center (VCDC), the requests that arrive from customers are dynamic. The applications that run in the VCDC generally require different types of resources. The VCDC will have to run a mix of applications of which some are CPU-intensive, while others are I/O-intensive. Thus, if we try to satisfy customer demand by carrying out the allocation process considering any one or two resources, much of the resources will be wasted. Bi et al. [2] propose an approach that is capable of handling applications that use different types of resources. A controller that can carry out resource allocation in a dynamic environment, while maintaining the SLA, has been proposed. The controller attempts to place applications that do not compete for the same type of resources, on the same node. In other words, applications that require the same type of resources will be placed on different nodes in order to avoid resource contention problems. The model developed by the authors attempt to balance the objectives of minimizing the energy cost and maximizing the profit of the VCDC. The proposed controller achieves its goal using the hybrid algorithm based on two metaheuristic algorithms—Particle Swarm Optimization (PSO) and Simulated Annealing (SA). Another aspect of this work is that they have proposed a model to calculate the rate of internal and external requests arriving at the VCDC. A probabilistic model to deal with nonsteady states has also been proposed. One of the notable characteristic of the proposed approach is that, it takes into consideration several factors such as the electricity price, which have not been considered in most approaches in the literature.

The Cloud environment is an environment that undergoes constant change. One of the main challenges in such a dynamic environment is the adaptation of the various resources in accordance with the changes that constantly occur. Bu et al. [4] attempt to tackle this problem by proposing an approach for how virtual machines and applications running on the virtual machines can respond to changes in the environment. The proposed framework CoTuner, applies a hybrid method that combines Simplex method and Reinforcement Learning (RL) Approach. Since the RL-Approach is used, the framework does not require any model. It does not require any dataset for training. The problem is formulated as a Markov decision problem. The CoTuner framework employs two types of agents: VM-Agents and App-Agents. VM-Agents are responsible for controlling the VMs that fall in the same VM cluster. The two levels of coordination are carried out repeatedly, one after the other, until the system reaches a balanced state. The framework relies on system knowledge obtained from two metrics of the system: the CPU Utilization and the memory utilization. Prototypes of the proposed model are developed and tested using TPCW and TPCC benchmark applications.

Energy consumption is one of the most crucial issues that is being considered by researchers worldwide, irrespective of the domain in which they are working in. The literature includes several such works by various researchers [3, 16, 19] which attempt to solve the problem of energy consumption. Some of these works propose the idea of sustainable data centers. One of the characteristics of the sustainable data centers is that a constant power supply cannot be guaranteed. This characteristic has been considered by Cheng et al. [8]. The power supply in sustainable data centers varies depending on the weather making it difficult for data centers to meet the demand of the customers with the available power supply. It is thus essential to develop a scheme that can find out the optimal resource allocation, taking into consideration the power supply, QoS level and other constraints. ePower derives the availability of resources from the current power supply. According to the availability, the resources can be allocated. The main idea is to delay the execution of batch jobs that take longer time and give a higher priority to transactional workloads. The automatic scheme proposed by the authors combines the abilities of a fuzzy model and simulated annealing. Using the ARIMA model, the power supply in the next interval is calculated based on the previous n observations.

The main characteristic of cloud computing that enables it to cater the needs of a large number of customers is its elasticity. At the same time, the elastic nature may also lead to some resources being wasted. In order to provide the elastic nature, many data centers allocate more resources than actually requested by the users, so that they can accommodate an increase in user request. This is called overbooking. If the overbooking is not done carefully, it will have an adverse effect on the cloud provider. Thus, overbooking is associated with its risks. Tomás et al. [24] propose an approach to calculate the risks involved which can aid the decision-making process of whether overbooking should be performed. In order to calculate the risk, fuzzy logic functions are utilized. The output of the fuzzy function helps in determining the amount of overbooking that can be done without negatively affecting the Cloud provider. Based on this decision, a service may be denied entry into the Data Center if the risk associated is calculated to be high. The acceptable level of risk is dynamically calculated using Proportional Integral Derivative (PID) Controller.

Cloud computing runs the various users' tasks by running their tasks on virtual machines. These virtual machines run on the physical machines. One of the major decision problems in Cloud computing is to decide on which of the physical machines, the virtual machines should be run. This decision problem is called the Virtual Machine Allocation Problem. The virtual machines provided to the users are supplied with enough resources as requested by the client or as required by the client applications. While deciding which physical machine can host each virtual machine, this requirement should be considered. After the initial allocation, in case of unexpected events such as load variations, this allocation will have to be changed by migrating VMs. Generally, data centers require the intervention of manual users in order to manage such events. Hyser et al. [13] propose an approach to automat-

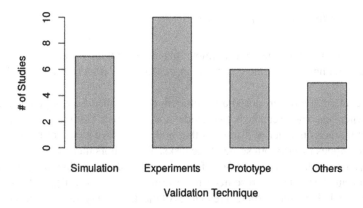

Fig. 1 Validation approaches employed by different research works

ically respond to such events. An autonomic controller that evaluates the current condition in the data center and develops the corresponding migration list has been proposed. A solver based on Simulated Annealing derives the solution to the optimization problem that has been formulated as an Iterative Rearrangement Problem (Fig. 1).

In order to support the elastic nature of Clouds, it is required to perform allocation of resources to applications. Gambi et al. [11] propose a solution based on Kriging models that perform this allocation by considering the QoS and execution costs. The controllers are made to be self-adaptive as the workload conditions vary greatly in the Cloud. The main strength of a Kriging model is that each prediction they make is associated with a level of confidence as well. The authors have applied this solution to solve the allocation problem at the IaaS layer. The goals of the optimization problem (i.e., reduce the probabilities of violations, minimize the amount of resources allotted, reduce the execution cost) are represented as a utility function. The authors have observed that such black-box models are better suited to the changing Cloud environment than white-box models.

Elasticity in the Cloud computing environment can be provided by performing autoscaling. For efficiently leveraging the autoscaling process, rules are required to guide the process of scaling up and scaling down. Generally, the thresholds are provided by the user. Jamshidi et al. [14] suggest that due to the unpredictable nature of the Cloud environment, there can be various ambiguities about the varying demand and other factors such as the time required to start a new VM. In order to bypass this, the authors suggest the use of a fuzzy system that can convert the rules into numerical values.

A controller is used to make the decisions on when and how to scale. The controller does this with the help of a fuzzy system. An online learning component based on Q-learning is also adapted in order to modify the rules for scaling in accordance with the changes in the system.

2.6 Miscellaneous Approaches

There are other approaches that employ various techniques. Some of these techniques use mathematical concepts. Such approaches and others are discussed in this subsection.

Che et al. [10] have proposed a novel architecture called Self-organizing Cloud (SoC), where each node can be both provider and consumer. Each of the nodes works in an autonomous manner to discover other nodes that have plenty of resources and uses those nodes to carry out some of the tasks. The authors then attempt to develop solutions for two design issues in SoC. First, they try to develop a language in order to communicate the request for resources along various dimensions. The node identified should be such that it has sufficient resources of the different types, as required by the task. Next, they try to derive an optimal allocation which can optimize the execution time. For this, first the authors prove the existence of an optimal solution using Slater's condition. Then, a convex optimization problem is formulated and solved using Proportional Share Model (PSM).

Viswanathan et al. [26] propose an approach that manages the energy efficiency of data centers. The proposed approach requires a knowledge of the thermal state of the data center. Within a data center, all the areas may not be at a same temperature. Owing to the heat generated by the servers in the data center and the action of the cooling equipment, some areas will be at a higher temperature than other areas. In order to know about the temperature in different areas of the data center, the proposed approach makes use of thermal cameras in order to capture thermal images. These thermal images are then analyzed in order to detect the hotspots within the data center so that focus can be on those areas in order to maintain the heat balance within the data center. The analyzed data can also be used to detect any abnormalities and take appropriate actions. The approach offers different levels of autonomy, starting from the sensors used to obtain the temperature information that can themselves decide the rate and region of sampling, ranging to the allocation and detection of abnormalities. The information obtained from analysis on the thermal image can be used to control Virtual Machine (VM) allocation in order to keep the heat generated in the data center within the bounded limits.

Mastroianni et al. [20] propose an approach, ecoCloud for improving the energy consumption in a self-organizing data center, where each server in the data center is an autonomous server. One of the widely used technologies to reduce the energy consumed by the data centers is virtual machine consolidation, where virtual machines are migrated to ensure that all the servers run at optimized loads. Any server with very less load will be powered off after the migrations have been completed. In most of the research works that adopt this technique, we generally have a managing entity taking the decisions of when and how the consolidation process should be carried out. The ecoCloud approach suggested by the authors, attempts to bring a change to the conventional trend by handing over the decision process to each of the autonomic servers in the Self-organizing Cloud. Each server can independently decide whether or not to accept a VM request, on the basis of the result of the Bernoulli trial performed. Similarly, when a VM running on the server is observed to have resource utilization (CPU or RAM) outside the bounds, a Bernoulli trial is performed to decide whether the migration process should be started or not. It is the central manager that has the final say in the overall control of the data center.

Virtual machine consolidation is a technique that is commonly employed in data centers in order to reduce energy consumption. While consolidating virtual machines, a crucial issue to be considered is the nature of the workloads that are being run on a particular physical machine. Often, the workloads are inherently different from each other, which poses further challenges in the management of execution. Different types of workloads have to be handled differently. Since all these workloads are co-located on the same host, the situation becomes even more complex. Carrera et al. [5] propose an approach that tries to reduce the complexity of the situation, by enabling the servers to estimate the level of performance of the jobs in the queue. In order to handle the mixed type of workloads seamlessly, the system uses Relative Performance functions which consider the performance level of each application with respect to the goal of the application. The placement controllers decisions are based on the value of this function. The system is continuously monitored to detect any changes, requiring a change in the current allocation.

Cloud bursting is a technology where a private cloud provider takes the help of a public cloud provider in order to service all of its clients' requests. In such scenarios, it is essential to determine where each job should be executed and when. Kailasam et al. [15] consider the execution of data-intensive jobs in such an environment. In the case of data-intensive jobs, the latency in shifting the data from one Cloud provider to the other must also be taken into consideration. The authors propose three heuristics which attempt to derive an optimal schedule for executing the jobs. The approach is composed of two phases. At the end of the first phase, the various outputs of the jobs will have to be ordered. The heuristics are analyzed based on three metrics specific to ordered throughput.

Zheng et al. [29] propose an approach to improve the reliability of Cloud systems by incorporating fault tolerance. Since the Cloud environment involves a large number of resources, it is nearly impossible to provide redundancy for all the units. Instead the critical components are identified and fault tolerance is provided for such components. To identify the critical components, the components are ranked based on the number of times it is being invoked by others. For this, the initial stage involves the construction of a component graph. After the critical components are identified, the optimal fault tolerance strategy is selected from a list of available strategies. This is done by considering the probability of failure of each strategy.

Cloud computing coupled with autonomicity can effectively cater to the needs of the users. Since the cloud environment involves a large number of resources, providing every resource with autonomic behavior may not be feasible at the initial stage. As the system evolves, it may be required that component not initially autonomic be made autonomic on the fly. Mohamed et al. [21] propose a framework that assists the process. The authors base their approach on the Open Cloud Computing Interface (OCCI) standard. Attributes that need to be monitored are identified and autonomic behavior is encorporated by using two OCCI components defined by the authors' links and mixins. The approach can be employed in any Cloud deployment model.

3 Discussion

In this study, we have categorized the various research works based on the methodology adopted. Table 1 gives a summary of the works considered in this study. We have found that there is active research going on in the field of autonomic computing. The self-*properties considered in the research works are summarized in Table 2. On analysis, we can see that most of the studies focus on any one self-*property. From Fig. 2, it is evident that self-configuration and self-optimization properties are considered by most of the researchers. It is also seen that some of the techniques focusing on one self-*property overlaps with other properties as well.

On the basis of our reading, we have attempted to classify the various works based on the methodology that they have used. Another way to classify them is by considering the self-*properties that each research work achieves. From Fig. 3, it can be seen that most of the researchers employ hybrid approaches which combines two or more methodologies. This can be attributed the fact that using one particular approach has its own strengths and limitations. By combining different approaches, the approaches complement each other, thus providing a better performance than either of the approach being employed individually.

Another aspect that we have considered in this study is the type of validation technique used by the various researchers. Figure 1 clearly shows that most of the researchers have performed experiments either on a testbed or in a real environment to validate their proposed approaches. Techniques supported by mathematical proofs are considerably less.

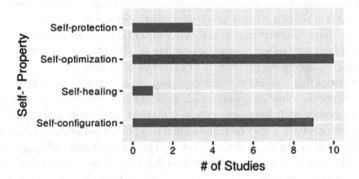

Fig. 2 Self-*properties considered in existing works

Table 1 Details of the existing works in literature

Study	Targetted domain	Problem	Objective	Methodology	Contribution	Validation approach
[2]	CPU and I/O	Resource allocation	Maximize profit of VCDC	Hybrid metaheuristic; PSO + SA	SLA of heterogeneous applications, # of rejected requests, price of electricity	Simulation (MATLAB, trace driven)
[4]	VMs and resident applications	Configuration	Automatic tuning of VM resource allocations and resident application parameter settings	Hybrid; simplex + RL	No model of system required	Prototype built
[8]	Resources of a data center	Optimal resource allocation	QoS requirements, dynamic green power supply	SA + fuzzy model	Non-static power supply considered to provision for heterogeneous workloads	Experiments
[24]	Resources of datacenter	Appropriate level of overbooking	Improve application performance without violating capacity constraints	Fuzzy logic functions + PID controller	Independent of user input for overbooking level	Simulations + experiments
[13]	VMs	Mapping of VMs to hosts	Follow policies specified by user	SA + control theory	Deal with unanticipated events without human intervention	Prototype
[11]	Resources in data center	Application elasticity	Fluctuating input workload	Kriging model + control theory	Requires no model	Prototype
[14]	Cloud resources	Autoscaling	Cope with uncertainties in Cloud	Fuzzy logic (Q-learning) + control theory	Considers fluctuating workloads	Experiments
[28]	Personal data of users	Protect user data privacy	Make data unreadable after a specified time	Cryptography + active storage principles	Distributed object-based storage system, user control over lifetime of private data	Prototype

(continued)

Table 1 (continued)

Study	Targetted domain	Problem	Objective	Methodology	Contribution	Validation approach
[27]	Users private data	Secure self-destructing scheme for data sharing	Fine-grained access control, sensitive data self-destruction	Key-policy attribute-based encryption	User-defined, time-specific authorization	Theoretical analysis ([expanded BDHI assumption])
[6]	QoS	Modeling QoS for software-based services	Dynamic, self-adaptive model	Information theory + machine learning	Dynamically select suitable ML algorithm for predicting correlation	Experiments
[30]	Inter-cloud resources	Task scheduling	Meet peak demand; maximize profit of provider; preserve QoS	Self-learning PSO	No formal agreement required between cloud providers	Experiments
[25]	Services	Clustering of similar services	Predict nature of incoming service	Random forest + partition around medoids	Can handle mixed data types, no model required	Experiments
[10]	Resources in a self-organizing cloud	VM-multiplexing resource allocation	Allocate node with available resource along all dimensions	Dynamic optimal proportional share using Proportional Share Model	Resources considered along multiple dimensions	Simulation (PeerSim)
[26]	Data center management	Communication and coordination for self-organization	Thermal aware operations	Analysis of thermal images	Cross-layered approach	Presented solution
[7]	Resources in data center	Autoscaling	Adaptive configuration and resource provisioning	Machine learning algorithms	QoS modeling based on online analysis	Presented solution
[17]	Services	Optimal service composition	QoS aware	GA	Considers QoS of network	Experiments
[23]	VMs	Prevent coresidence profiling	Cluster users and workloads according to preferences, obfuscate IP map	Bio-inspired metaheuristic (mussel-inspired)	Mathematical model for mussel interactions	Risk assessment
[22]	Services	Recomposition of services	QoS requirements	Market-based Continuous Double Auction	Focus on good solution rather than optimal solution	Simulation

(continued)

Table 1 (continued)

Study	Targetted domain	Problem	Objective	Methodology	Contribution	Validation approach
[20]	VMs	VM consolidation	CPU and RAM requirements	Probabilistic model: Bernoulli trial	More than one resource dimension considered	Experiments + mathematical model
[18]	Cloud resources in distributed cloud	Resource allocation	Application requirements	Market-based Vickrey auctions	Scalable, computationally tractable	Simulation
[5]	Workloads	Placement of heterogeneous workload	Maximize global objective of system	Virtualization control mechanisms + relative performance function	Considers mixed type of workloads, application-centric, maximizes performance of least performing application in system	Simulation + prototype
[15]	Data-intensive applications	Optimize ordered throughput	Inter-cloud bandwidth, large data transfer requirements	Scheduling heuristics	Considers download time, considers online task arrival	Experiment + simulation
[29]	Cloud application components	Optimal fault tolerance strategy	Response time, cost	Component ranking based	Systematic ranking-based framework	Prototype
[21]	Cloud resources	Describe resources that need autonomic capability	Add autonomic management capability	Model based on Petri-nets	Generic solution applicable to all layers	Experiments

Table 2 Self-*properties of the studies

Study	self-configuring	self-healing	self-optimizing	self-protecting
[2]	×	×	3	×
[4]	3	×	×	×
[6]	3	×	×	×
[8]	×	×	3	×
[30]	3	×	×	×
[10]	×	×	3	×
[28]	×	×	×	3
[27]	×	×	×	3
[26]	×	×	3	×
[24]	3	×	×	×
[7]	3	×	×	×
[13]	×	×	3	×
[17]	×	×	3	×
[23]	×	×	×	3
[22]	3	×	×	×
[11]	×	×	3	×
[20]	×	×	3	×
[18]	3	×	×	×
[5]	×	×	3	×
[14]	3	×	×	×
[15]	3	×	3	×
[29]	×	3	×	×

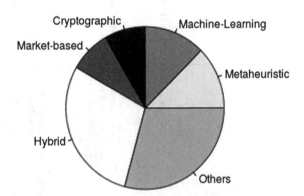

Fig. 3 Methodologies used by researchers to achieve autonomicities

4 Future Directions

On the basis of our study, we have observed that:

– Currently, there is a lack of methods to compare the performances of various autonomic cloud controllers. As the number of approaches to provide autonomicity increases, researchers will also have to focus on metrics that can be used to rate the performance of the controller.
– Though autonomic computing envisions the holistic development of the system, focus has to be directed to derive the optimal degree of autonomicity in Cloud. There are different degrees of autonomicities ranging from fully manual to fully automated. The trade-offs should be considered to derive an optimal degree of autonomicity in Cloud environments.
– In real-time scenarios, Cloud environments are initially built without autonomic capabilities and on the long run, encorporated with self-managing powers. The overhead involved in incorporating autonomic functionalities in a real environment needs to be considered.
– There are only a few approaches that realize the self-healing property in Autonomic Clouds. Self-healing property can be provided to an Autonomic Cloud by providing fault tolerance and improving the reliability of the systems. Researchers can attempt to make their systems capable of healing themselves, which includes the process of detecting and reacting to the faults.
– The researchers generally focus on one self-*property. Systems employing a combination of the self-*properties are required.

5 Conclusion

In this study, we have considered the various works in literature that are directed toward the autonomic management of resources in Cloud Computing. The various works are categorized based on the methodology that they have used. An analysis of the different aspects of the research works is done. Based on the results of the analysis, some future directions have been proposed.

From the study, it is evident that research in the field of autonomic Clouds has great potential to be explored further. The level of autonomicity is very low in the current Cloud environments. Efforts have to be put in to increase the level of autonomicity. But, this has its own trade-offs. So, the first step for researchers is to determine which is the optimal level of autonomicity. Further, while building autonomic systems, more focus has to be given to the self-healing capability.

References

1. Berns, A., Ghosh, S.: Dissecting self-*properties. In: 2009 Third IEEE International Conference on Self-Adaptive and Self-Organizing Systems, pp. 10–19. IEEE (2009)
2. Bi, J., Yuan, H., Tan, W., Zhou, M., Fan, Y., Zhang, J., Li, J.: Application-aware dynamic fine-grained resource provisioning in a virtualized cloud data center
3. Bilal, K., Malik, S.U.R., Khalid, O., Hameed, A., Alvarez, E., Wijaysekara, V., Irfan, R., Shrestha, S., Dwivedy, D., Ali, M., et al.: A taxonomy and survey on green data center networks. Future Gener. Comput. Syst. **36**, 189–208 (2014)
4. Bu, X., Rao, J., Xu, C.Z.: Coordinated self-configuration of virtual machines and appliances using a model-free learning approach. IEEE Trans. Parallel Distrib. Syst. **24**(4), 681–690 (2013)
5. Carrera, D., Steinder, M., Whalley, I., Torres, J., Ayguade, E.: Autonomic placement of mixed batch and transactional workloads. IEEE Trans. Parallel Distrib. Syst. **23**(2), 219–231 (2012)
6. Chen, T., Bahsoon, R.: Self-adaptive and online qos modeling for cloud-based software services. IEEE Trans. Softw. Eng. (2015)
7. Chen, T., Bahsoon, R.: Toward a smarter cloud: self-aware autoscaling of cloud configurations and resources (2015)
8. Cheng, D., Rao, J., Jiang, C., Zhou, X.: Elastic power-aware resource provisioning of heterogeneous workloads in self-sustainable datacenters. IEEE Trans. Comput. **65**(2), 508–521 (2016)
9. Computing, A.: An Architectural Blueprint for Autonomic Computing. IBM Publication (2003)
10. Di, S., Wang, C.L.: Dynamic optimization of multiattribute resource allocation in self-organizing clouds. IEEE Trans. Parallel Distrib. Syst. **24**(3), 464–478 (2013)
11. Gambi, A., Pezze, M., Toffetti, G.: Kriging-based self-adaptive cloud controllers. IEEE Tran. Serv. Comput. **99**, 1–1 (2015)
12. Horn, P., Computing, A.: IBMs Perspective on the State of Information Technology. IBM TJ Watson Labs, NY, 15 October (2001)
13. Hyser, C., Mckee, B., Gardner, R., Watson, B.J.: Autonomic virtual machine placement in the datacenter. Hewlett Packard Laboratories, Tech. Rep. HPL-2007-189 **189** (2007)
14. Jamshidi, P., Pahl, C., et al.: Managing uncertainty in autonomic cloud elasticity controllers. IEEE Cloud Comput. **3**(3), 50–60 (2016)
15. Kailasam, S., Gnanasambandam, N., Dharanipragada, J., Sharma, N.: Optimizing ordered throughput using autonomic cloud bursting schedulers. IEEE Trans. Softw. Eng. **39**(11), 1564–1581 (2013)
16. Kaur, T., Chana, I.: Energy efficiency techniques in cloud computing: a survey and taxonomy. ACM Comput. Surv. (CSUR) **48**(2), 22 (2015)
17. Klein, A., Ishikawa, F., Honiden, S.: Sanga: a self-adaptive network-aware approach to service composition. IEEE Trans. Serv. Comput. **7**(3), 452–464 (2014)
18. Landa, R., Charalambides, M., Clegg, R.G., Griffin D., and Rio, M.: Self-Tuning Service Provisioning for Decentralized Cloud Applications. In: IEEE Transactions on Network and Service Management, vol. 13, no. 2, pp. 197–211, June 2016. http://dx.doi.org/10.1109/TNSM.2016.2549698
19. Liu, L., Wang, H., Liu, X., Jin, X., He, W.B., Wang, Q.B., Chen, Y.: Greencloud: a new architecture for green data center. In: Proceedings of the 6th International Conference Industry Session on Autonomic Computing and Communications Industry Session. pp. 29–38. ACM (2009)
20. Mastroianni, C., Meo, M., Papuzzo, G.: Probabilistic consolidation of virtual machines in self-organizing cloud data centers. IEEE Trans. Cloud Comput. **1**(2), 215–228 (2013)
21. Mohamed, M., Amziani, M., Belaïd, D., Tata, S., Melliti, T.: An autonomic approach to manage elasticity of business processes in the cloud. Future Gener. Comput. Syst. **50**, 49–61 (2015)
22. Nallur, V., Bahsoon, R.: A decentralized self-adaptation mechanism for service-based applications in the cloud. IEEE Trans. Softw. Eng. **39**(5), 591–612 (2013)

23. Rice, J.L., Phoha, V.V., Robinson, P.: Using mussel-inspired self-organization and account proxies toobfuscate workload ownership and placement in clouds. IEEE Trans. Inf. Forensics Secur. **8**(6), 963–972 (2013)
24. Tomás, L.: Tordsson J (2014) An autonomic approach to risk-aware data center overbooking. IEEE Trans. Cloud Comput. **2**(3), 292–305 (2014)
25. Uriarte, R.B., Tsaftaris, S., Tiezzi, F.: Supporting autonomic management of clouds: service clustering with random forest
26. Viswanathan, H., Lee, E.K., Pompili, D.: Self-organizing sensing infrastructure for autonomic management of green datacenters. IEEE Network **25**(4), 34–40 (2011)
27. Xiong, J., Liu, X., Yao, Z., Ma, J., Li, Q., Geng, K., Chen, P.S.: A secure data self-destructing scheme in cloud computing. IEEE Trans. Cloud Comput **2**(4), 448–458 (2014)
28. Zeng, L., Chen, S., Wei, Q., Feng, D.: Sedas: A self-destructing data system based on active storageframework. In: APMRC, 2012 Digest. pp. 1–8. IEEE (2012)
29. Zheng, Z., Zhou, T.C., Lyu, M.R., King, I.: Ftcloud: A component ranking framework for fault-tolerantcloud applications. In: 2010 IEEE 21st International Symposium on Software Reliability Engineering. pp. 398–407. IEEE (2010)
30. Zuo, X., Zhang, G., Tan, W.: Self-adaptive learning pso-based deadline constrained task scheduling for hybrid Iaas cloud. IEEE Trans. Autom. Sci. Eng. **11**(2), 564–573 (2014)

Toward Efficient Autonomic Management of Clouds: A CDS-Based Hierarchical Approach

John Paul Martin, A. Kandasamy and K. Chandrasekaran

Abstract Cloud computing is one of the most sought-after technologies today. Beyond a shadow of doubt, the number of clients opting for Cloud is increasing. This steers the complexity of the management of the Cloud computing environment. In order to serve the demands of customers, Cloud providers are resorting to more resources. Relying on a single managing element to coordinate the entire pool of resources is no more an efficient solution. Therefore, we propose to use a hierarchical approach for autonomic management. The problem that we consider here is to determine the nodes at which we have to place the Autonomic Managers (AMs), in order to ease the management process and minimize the cost of communication between the AMs. We propose a graph-theory-based model using Connected Dominating Set (CDS) that allows to determine an effective placement of AMs in different Data Centers (DCs), and, their collaboration with the Global Manager (GM). The approach considers the construction of domination sets and then, distributing the control of the dominees among the dominators.

1 Introduction

Cloud Computing can be visualized as dynamically scalable shared resources accessed over a network. Basically, cloud supports three platforms: Infrastructure as a Service, Platform as a Service, and Software as a Service. Nowadays, apart

J. P. Martin (✉) · A. Kandasamy
Department of Mathematical and Computational Sciences, National Institute of Technology Karnataka, Surathkal 575025, India
e-mail: johnpm12@gmail.com

A. Kandasamy
e-mail: kandy@nitk.ac.in

K. Chandrasekaran
Department of Computer Science and Engineering, National Institute of Technology Karnataka, Surathkal 575025, India
e-mail: kchnitk@gmail.com

© Springer Nature Singapore Pte Ltd. 2018
R. Chaki et al. (eds.), *Advanced Computing and Systems for Security*, Advances in Intelligent Systems and Computing 667,
https://doi.org/10.1007/978-981-10-8183-5_4

from this, various cloud providers such as IBM [1], Amazon, HP, Google, Citrix, Salesforce, etc., are providing a wide variety of services which ranges from normal text editing softwares to online healthcare services. Thus, we can say that, now, Cloud provides X as a Service (XaaS). The advantage of Cloud is that while choosing from a wide variety of services offered, user requires to pay only for the resources used. The demand for cloud users is increasing rapidly. People are also looking for ways to make the Cloud better fit their personal or organizational needs. Cloud providers should be able to solve the resource requirements with greater operational efficiencies. For meeting all the needs of the clients, cloud providers should support scalability. The providers should have the ability to solve fluctuations in business size and needs of the users. They should be able to scale up or down corresponding to the needs of the clients. Providing scalability is simultaneously the greatest advantage and challenges faced by cloud providers.

Using just a few data centers, providers cannot fulfill all the requirements of the clients. Cloud providers generally require data centers in the range of 100 s to support the clients of the Cloud. To guarantee performance, it is essential that these data centers are distributed across the globe rather than being concentrated in a particular region.

Nowadays, various cloud orchestration tools provide mechanisms to form the cloud by combining various data centers of the same vendor or different vendors to increase the scalability. The term federated cloud is used for interconnecting two or more computing environments of different vendors for meeting the spikes in demand. OpenStack provides a cascading solution for integrating multi-site OpenStack clouds. This provides new features such as the ability to combine 100 K servers, 100 data centers, and can run up to one million virtual machines. It also supports dynamically adding new data centers [2]. CloudStack uses the term 'region' [3], which is formed by combining two or more data centers which are referred to as zones [4].

Graph theory is often employed to solve network-related problems. We propose to use the concept of Connected Dominating Set (CDS), which can be used for the effective placement of Autonomic Managers in the Cloud infrastructure. A CDS-based hierarchical approach can reduce the communication overhead [5], avoid redundant information, and thereby provide a platform for effective management of the entire infrastructure.

The rest of this paper is structured as follows: Sect. 2 describes the motivation for this work and Sect. 3 contains the related works in this area and our proposed framework is explained in detail in Sect. 4. Discussions related to the proposed method are in Sects. 5 and 6 gives the conclusion.

2 Motivation

The International Data Corporation (IDC) predicted that the growth rate of cloud services will be approximately six times the rate of the overall IT growth. By 2019, the turnover from this business will be around $141 billion [6].

Manual management of a large-scale, complex structure like the Cloud is a tedious task, because of its heterogeneity in resources, dynamic infrastructure, and unexpected failures. An efficient managing system must be able to tackle the inherent challenges of controlling the geographically distributed centers. For the management of such a distributed system, it is suggested to be autonomic in nature [7–9]. Autonomic systems are self-configuring, self-healing, self-optimizing, and self-protecting [10]. Autonomic cloud systems contain Autonomic Managers for achieving these properties. One Autonomic Manager is not sufficient to control the entire infrastructure, multiple Autonomic Managers are required [11]. On the other hand, placing Autonomic Managers at every node will cause an exponential increase in the communication cost [12] and may also result in redundancy.

The management of such fast-growing complex infrastructure can be done effectively by making it autonomic in nature. Autonomic Managers or Autonomic Controllers are used to provide autonomicity in the system. A single AM is not sufficient to control the entire system. Placing of Autonomic Managers on all the nodes will be costly. Thus, there is a trade-off to be handled in the placing of Autonomic Managers. In this paper, we propose an approach that attempts to reduce the complexity of managing a distributed environment.

3 Related Work

There are many works by researchers in the field of autonomic computing. Some of the relevant works are mentioned in this section.

Casalicchio et al. [13] presented a model for the design of a self-optimizing Cloud Provider. The method will increase the total revenue of the provider by considering the available resources, SLA and constraints in the VM migration. They used the hill climbing algorithm for getting a near-optimal result and compared the obtained results with a Best Fit (BF) strategy. Mohamed and Megahed [12] proposed a mathematical model which determines the number of Autonomic Managers required for the effective management of a system. Their proposed model uses integer programming for deciding the number of Autonomic Managers.

Diaz-Montes et al. [14] proposed a federation model for the autonomic workflow executions. They made a detailed analysis of their model in Molecular dynamics simulations, Business Data Analytics, and Medical image research, which does not deal with where to place these Autonomic Managers.

Uriarte et al. [15] used unsupervised random forest algorithm to calculate the similarities among services and send the results to a clustering algorithm. This helps

Table 1 Summary of graph theory techniques used in the literature

No.	Problem addressed	Graph theory technique	Self-capability	Deployment environment
1	Task assignment [22]	Hypergraph	×	Cloud
2	Resource allocation [23]	Matching	×	Cloud
3	MapReduce applications [24]	Flow-network-based algorithm	×	Cloud
4	Service reliability [25]	Spanning tree	×	Cloud
5	Video distribution [26]	Dominating set	×	Cloud
6	Social networking [27]	Degree centrality	×	Cloud
7	Topology [28]	Topological graph theory	×	Cloud
8	Security [29, 30]	Associative bipartite graphs	×	Social networks
9	Trust management [31]	Algebraic graph theory	Self-organizing	Wireless network
10	Model the cloud network [32]	Predicate-based graph	×	Cloud

to make the Autonomic Manager aware about the similarities among services and their clusters and reduces the efforts required by Autonomic Manager.

Table 1 summarizes the graph theory concepts adopted by some of the relevant research works in the literature. There are a very few number of approaches that use the concept of Dominating Set. Among them, none of the approaches focus on equipping the system with self-managing capabilities. CDS has been used in different applications at various levels in wireless sensor networks [16, 17], but very few in Cloud.

Our approach combines the concepts of Connected Dominating Set and autonomic management of the Cloud.

4 Proposed Framework

4.1 Preliminaries

Definition 1 *Dominating set:* Dominating Set D, for a graph G with V vertices and E edges can be defined as a subset of V, where every vertex of V which are not in D are adjacent to at least one element of D with one hop distance.

Definition 2 *Connected Dominating Set:* Connected Dominating Set C, for a graph G with V vertices and E edges has the following properties:

- Subgraph C is a Dominating Set
- Subgraph C is connected in nature, that is there exists a path between any two nodes in C.

4.2 Details of the Proposed Framework

We propose a framework for placing Autonomic Managers in large-scale distributed systems like Cloud. The proposed framework uses a hierarchical approach that adopts the concept of Connected Dominating Set from graph theory for the placement of Autonomic Managers and thus helps in the efficient provisioning of resources and scheduling of applications.

A large Cloud infrastructure consists of many data centers. Each of these data centers contains many physical machines (also termed as hosts). The entire cloud infrastructure is managed by a Global Manager (GM). Clients can submit their jobs to the interface provided by the GM. The GM will be aware of the various data centers that are geographically separated in that cloud. GM collects this information by interacting with the various Region Managers (RM) in each region. A region is a collection of data centers that are decided based on the Connected Dominating Set (CDS). The RM gets information about each data center from the Host Group Managers (HGM). The GM decides where a particular job can be executed (Figs. 1 and 2).

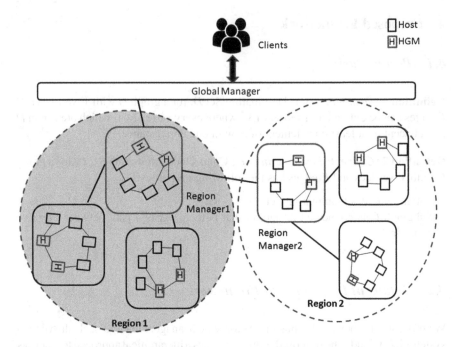

Fig. 1 Hierarchical CDS-based framework: Clients submit jobs to the cloud via the cloud interface, which is passed to the Global Manager. The Global Manager communicates with the Region Managers (represented by red-outlined DCs), who in turn communicate with the Host Group Managers (red-bordered hosts)

Algorithm 1 Host Group Manager (HGM) Operations

A Data Center(DC) can be represented as a graph G.

Let V be the set of vertices (Hosts), E be the set of edges(connections between hosts) in the graph.

Let $V_1, V_2 .... V_n$ be the nodes in V and v be any arbitrary node in V.

Let $deg(v)$ be the degree of node v and $n(v)$ be the neighbor set of any node v.

1 Find out the elements of CDS using Algorithm 3;
2 Place autonomic managers at each of the CDS nodes;
3 Select Host Group Manager among the autonomic managers;
4 Select node u with maximum $m(u)$ as Host Group Manager(HGM); where,

$$m(u) = \begin{cases} number\ of\ black\ color\ neighbours\ of\ u, & \text{if } u \in CDS \\ 0, & u \notin CDS \end{cases}$$

5 HGM collects information of all the hosts in the datacenter through the respective AMs and passes this information to the Region Manager;
6 Information obtained from the region manager is communicated to the AMs within the DC by the HGM;

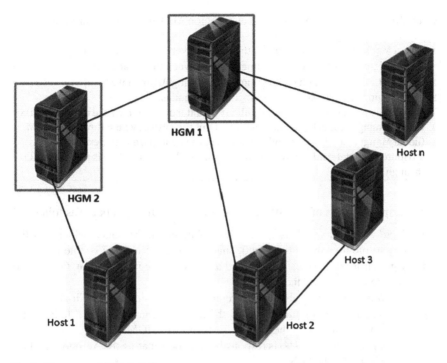

Fig. 2 Placement of AMs in DC

In our proposed approach, instead of placing autonomic controllers at every node in the data center, we place them at the nodes that are vertices of the CDS obtained from the graph obtained by considering the data center network as a graph. Each data center can be considered as a graph, where vertices are the hosts in the DC and edges are the connections between them. Among these autonomic controllers, a coordinator for each data center is selected using Algorithm 1. This coordinator is called Host Group Manager. At the next level, another graph is constructed with the different data centers as the nodes and the connections between them as edges. The CDS for this graph is also constructed. The HGMs at the data centers that form the nodes of the CDS are selected as Region Managers. The Region Managers are responsible for controlling the region that falls under it and also handles communication with the Global Manager (GM). The data centers that fall within each region are decided using Algorithm 2.

Algorithm 2 Global Manager (GM) and Region Manager (RM) Operations

1 Global Manager (GM) considers all the HGMs as vertices and constructs a graph, where edges E are the existing connections between the datacenters; 2 Find the CDS of the graph connecting the DCs using Algorithm 3; 3 Place Region Managers(RM) at each of the CDS nodes;   Each RM is responsible for the neighbouring datacenters which are not in the CDS; 4 Global Manager collects the information about the entire system through the RM; 5 Global Mangers analyze the collected information compare with performance attributes and determine any changes required. The Change Plan is passed to DCs through the RMs.

The responsibilities of the different managers in our framework are as follows:

- A *Global Manager (GM)* is the top-level Autonomic Manager, which has the capacity to monitor the entire cloud, analyze the collected and is able to decide if any global-level changes are required and coordinate its execution. GM is also responsible for selecting the RMs.
- *Region Manager (RM)* is able to take region-level decisions in the Cloud. Region Managers carries out all interactions with the Data Centers (DC) that falls within the concerned region, with the help of the respective HGMs.
- *Host Group Manager (HGM)* is responsible for coordinating the Autonomic Managers (AM) within the DCs.

There are various algorithms proposed for the construction of CDS [18]. We use the centralized CDS construction approach proposed by Guha and Khuller [19, 20]. Algorithm 3 gives the steps for CDS Construction.

5   Results and Discussion

In this paper, we proposed an efficient framework for the autonomic management of the large cloud infrastructure. Many of the researchers have proposed centralized approaches for autonomic management while some other approaches are based on distributing the management activity to all the nodes. Both the methods have its own demerits. We attempt to manage this trade-off by proposing a novel hierarchical approach which uses the concept of Connected Dominating Set in Graph Theory for distributing the work across various levels of Autonomic Managers, rather than all the nodes in the data centers. Our research focused on how we can effectively distribute the management activities across various levels and the collaboration between these levels. In order to validate our method, we created different networks using OPNET [21]. We consider two scenarios: the conventional scenario, where all nodes directly access services and the proposed approach where services are provided by the dominating nodes. Initially, we considered a network of 100 nodes. Later, we scaled it to 200 nodes and compared the performance metrics.

Algorithm 3 Algorithm for CDS Construction

Let V be the set of vertices and E be the set of edges in graph G.
Let C be an empty set.
Let $V_1, V_2 .... V_n$ be the nodes in V and v be any arbitrary node in V.
Let $deg(v)$ be the degree of node v and $n(v)$ be the set of 1-hop neighbours of any
   node v.
Data: Graph G
Result: Connected Dominating Set C
1 **for** *every node $v \in V$* **do**
2   colour[v]=white;
3 **end**
4 **for** *each $v \in V$* **do**
5   find deg(u);
6   **end**
7   Find $v \in V$ with maximum of $deg(v)$ and colour[v]= white; 8 **do**
colour[v]=black;
9   **for** *every node $y \in n(v)$* **do**
10   colour[y]=gray;
11   **end**
12   **while** *colour of any node $\in V$ equal to white* **do**
13   T=(Select v_i such that colour[v_i]=gray **or** (select $v_i, v_j \in n(v_i)$ where
    colour[v_i]=gray and color[v_j]=white) such that T has maximum number of
    incident edges;
14   **do** colour[T]=black;
15   **for** *every node $y \in n(T)$* **do**
16   colour[y]=gray
17   **end**
18   **end**
19   Connected Dominating Set C = Set of all nodes in G where colour[G]=black;

A detailed analysis of the results shows that the average CPU utilization (Fig. 3a, b) and delay (Fig. 4a, b) in the proposed approach is lesser than that in the conventional approach and it preserves that nature, on scaling as well. On comparing the utilization and delay values in the scaled (Figs. 3b and 4b) and nonscaled scenario (Figs. 3a and 4a), we find that the variation is lower for the proposed approach. This indicates that the proposed approach offers scalability as well.

Our framework can handle changes in the deployment level such as failure of nodes and addition of new nodes. These changes can be handled by the managers at the immediate higher level, without affecting the performance of the system. For example, if we want to add a host in a data center, the HGM can allot the new node to the corresponding Autonomic Manager by reconstructing the CDS. This change will not affect the nodes at the higher levels. On the other hand, in case of a component failure, the corresponding HGM will be notified and it can take further actions based on its knowledge of the data center actions, without the intervention of the GM.

When we are using the concept of Connected Dominating Set for finding AM and RM there is a chance that a particular node may fall into more than one region. Consider the network in Fig. 5. Here, there is an ambiguity as one node (Host 2 in Fig. 5) is the neighboring element of more than one Autonomic Manager (AM1 and AM2). In such cases, the manager at the next level (here HGM), has the ability to

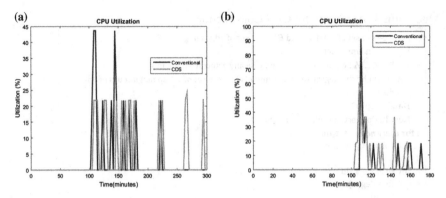

Fig. 3 CPU utilization values for **a** small-scale network and **b** large-scale network

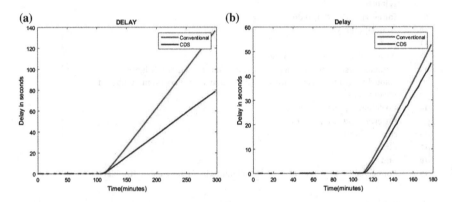

Fig. 4 Plot of delay for **a** small-scale network and **b** large-scale network

decide which Autonomic Manager should manage that node. It resolves this issue by taking the number of nodes that are present in each of the regions. The node will be assigned to the manager with a lesser number of nodes to control (here, AM2).

6   Conclusion

The management of the numerous resources in Cloud environment demands the use of managers with autonomic capabilities, so as to deal with the dynamically changing environment. Beholding this requirement, we proposed a hierarchical organization of AMs to manage the cloud resources. We use a graph-theory-based approach using the concept of Connected Dominating Set. The approach proposes a possible layout of the nodes at which we have to place the AMs in order to ease the management process of the distributed data centers and minimize the cost of communication between the

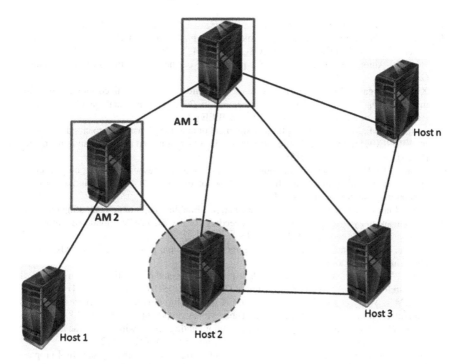

Fig. 5 Ambiguity in DC

AMs. Employing a hierarchical distribution of Autonomic Managers ensures that the nodes get serviced from the nearest layer, thereby to reducing energy consumption as well.

References

1. Smart cloud. http://www.ibm.com/cloud-computing/
2. Openstackcascading.   https://www.openstack.org/videos/video/building-multisite-multi-openstack-cloud-with-openstack-cascading
3. Apache cloudstack documentation. http://docs.cloudstack.apache.org/en/latest/concepts.html
4. Martin, J.P., Hareesh, M., Babu, A., Cherian, S., Sastri, Y., et al.: Learning environment as a service (leaas): cloud. In: 2014 Fourth International Conference on Advances in Computing and Communications (ICACC), pp. 218–222. IEEE (2014)
5. de Oliveira, F.A., Ledoux, T., Sharrock, R.: A framework for the coordination of multiple autonomic managers in cloud environments. In: 2013 IEEE 7th International Conference on Self-adaptive and Self-organizing Systems, pp. 179–188. IEEE (2013)
6. Worldwide public cloud services spending forecast to double by 2019, according toidc. https://www.idc.com/getdoc.jsp?containerId=prUS40960516
7. Buyya, R., Calheiros, R.N., Li, X.: Autonomic cloud computing: open challenges and architectural elements. In: 2012 Third International Conference on Emerging Applications of Information Technology (EAIT), pp. 3–10. IEEE (2012)

8. Kephart, J.O., Chess, D.M.: The vision of autonomic computing. Computer **36**(1), 41–50 (2003)
9. Singh, S., Chana, I.: Qos-aware autonomic resource management in cloud computing: a systematic review. ACM Comput. Surv. (CSUR) **48**(3), 42 (2016)
10. Horn, P.: Autonomic computing: Ibm\'s perspective on the state of information technology (2001)
11. Xia, Y., Tsugawa, M., Fortes, J.A., Chen, S.: Toward hierarchical mixed integer programming for pack-to-swad placement in datacenters. In: 2015 IEEE International Conference on Autonomic Computing (ICAC), pp. 219–222. IEEE (2015)
12. Mohamed, M., Megahed, A.: Optimal assignment of autonomic managers to cloud resources. In: 2015 IEEE International Conference on Service Operations and Logistics, and Informatics (SOLI), pp. 88–93. IEEE (2015)
13. Casalicchio, E., Menascé, D.A., Aldhalaan, A.: Autonomic resource provisioning in cloud systems with availability goals. In: Proceedings of the 2013 ACM Cloud and Autonomic Computing Conference, p. 1. ACM (2013)
14. Diaz-Montes, J., Zou, M., Rodero, I., Parashar, M.: Enabling autonomic computing on federated advanced cyberinfrastructures. In: Proceedings of the 2013 ACM Cloud and Autonomic Computing Conference, p. 20. ACM (2013)
15. Uriarte, R.B., Tsaftaris, S., Tiezzi, F.: Supporting autonomic management of clouds: service clustering with random forest
16. Dai, F., Wu, J.: An extended localized algorithm for connected dominating set formation in ad hoc wireless networks. IEEE Trans. Parallel Distrib. Syst. **15**(10), 908–920 (2004)
17. Thai, M.T., Wang, F., Liu, D., Zhu, S., Du, D.Z.: Connected dominating sets in wireless networks with different transmission ranges. IEEE Trans. Mob. Comput. **6**(7), 721–730 (2007)
18. Yu, J., Wang, N., Wang, G., Yu, D.: Connected dominating sets in wireless ad hoc and sensor networks—a comprehensive survey. Comput. Commun. **36**(2), 121–134 (2013)
19. Guha, S., Khuller, S.: Approximation algorithms for connected dominating sets. In: European Symposium on Algorithms, pp. 179–193. Springer, Berlin (1996)
20. Guha, S., Khuller, S.: Approximation algorithms for connected dominating sets. Algorithmica **20**(4), 374–387 (1998)
21. Opnet. https://www.riverbed.com/in/products/steelcentral/opnet.html?redirect=opnet
22. Çatalyürek, U.V., Kaya, K., Uçar, B.: Integrated data placement and task assignment for scientific workflows in clouds. In: Proceedings of the Fourth International Workshop on Data-Intensive Distributed Computing, pp. 45–54. ACM (2011)
23. Bansal, N., Lee, K.W., Nagarajan, V., Zafer, M.: Minimum congestion mapping in a cloud. In: Proceedings of the 30th Annual ACM SIGACT-SIGOPS Symposium on Principles of Distributed Computing, pp. 267–276. ACM (2011)
24. Li, M., Subhraveti, D., Butt, A.R., Khasymski, A., Sarkar, P.: Cam: a topology aware minimum cost flow based resource manager for MapReduce applications in the cloud. In: Proceedings of the 21st International Symposium on High-Performance Parallel and Distributed Computing, pp. 211–222. ACM (2012)
25. Dai, Y.S., Yang, B., Dongarra, J., Zhang, G.: Cloud service reliability: modeling and analysis. In: 15th IEEE Pacific Rim International Symposium on Dependable Computing, pp. 1–17. Citeseer (2009)
26. Sobiya, P., Nayagam, M.G.: Dominating set based content cloud architecture for video distribution services. In: 2014 International Conference on Green Computing Communication and Electrical Engineering (ICGCCEE), pp. 1–6. IEEE (2014)
27. VijayaChandra, J., Rao, K.T., Reddy, V.K.: Numerical formulation and simulation of social networks using graph theory on social cloud platform. Glob. J. Pure Appl. Math. **11**(3), 1253–1264 (2015)
28. Binz, T., Fehling, C., Leymann, F., Nowak, A., Schumm, D.: Formalizing the cloud through enterprise topology graphs. In: 2012 IEEE 5th International Conference on Cloud Computing (CLOUD), pp. 742–749. IEEE (2012)
29. Burgess, M., Canright, G., Engø-Monsen, K.: A graph-theoretical model of computer security. Int. J. Inf. Secur. **3**(2), 70–85 (2004)

30. Zegzhda, P.D., Zegzhda, D.P., Nikolskiy, A.V.: Using graph theory for cloud system security modeling. In: International Conference on Mathematical Methods, Models, and Architectures for Computer Network Security, pp. 309–318. Springer, Berlin (2012)
31. Jiang, T., Baras, J.S.: Graph algebraic interpretation of trust establishment in autonomic networks. Prep. Wiley J. Netw. (2009)
32. Chan, W., Mei, L., Zhang, Z.: Modeling and testing of cloud applications. In: APSCC2009, pp. 111–118 (2009)

Delta Authorization Concept for Dynamic Access Control Model in Cloud Environment

Sayantani Saha, Rounak Das and Sarmistha Neogy

Abstract Advancement in distributed computing and Internet-based computing, like cloud, has put high concerns on security issues. Identity and access management is one such issue that requires urgent attention. Both data privacy and user privacy need to be protected in privacy-aware cloud computing applications. Anonymous user interaction helps users to privately interact with any system. It must be made sure that unauthorized entity should not get access to data resources. Hence, identity credentials may not always be enough. Different contexts like user role, trust, behavior may be considered as an identity context required to authenticate the user for an active session. In any privacy-aware system, a wrong interaction might provide improper data access. Therefore, dynamic decision-making may also be required for a proper access session to continue. Quasi-static authorization models consider reauthorization at regular intervals. At the end of each interval, the user is again verified against his various contexts if the resource access is to be continued. Many research works focus on fine-grained access control model in highly dynamic environments. However, handling the contexts is the main concern in these types of access model. In this chapter, we propose an access model with the concept of delta authorization with an easy user session out process. Here, authorization not only proceeds at delta intervals, but also considers contextual information besides usual credentials. This technique is able to handle fine-grained access control in a better way. An implementation with the analysis is also presented.

Keywords Authentication · Authorization · Dynamic context · Quasi-static Authorization · Dynamic authorization · Access model

S. Saha (✉)
Information Technology, B P Poddar Institute of Management & Technology, Kolkata, India
e-mail: sayantanircc@gmail.com

R. Das
Adobe India, Chennai, India

S. Neogy
Department of Computer Science & Engineering, Jadavpur University, Kolkata, India

© Springer Nature Singapore Pte Ltd. 2018
R. Chaki et al. (eds.), *Advanced Computing and Systems for Security*, Advances in Intelligent Systems and Computing 667,
https://doi.org/10.1007/978-981-10-8183-5_5

1 Introduction

Authentication and authorization are two integral parts of an identity and access management model. These processes work in tandem to provide secured request execution. The process of authentication validates the subject (user) credential, whereas authorization decides the valid set of actions of the subject on the specific object (resource). Different authentication and authorization factors ultimately lead to valid resource access. Any context information change may invalidate the user identity and consequently, the authorization too gets restricted. So the simple concept of user validation, i.e., user authentication by providing id/password for necessary access permissions can no longer be considered sufficient in a dynamic environment like cloud. Such a system may not always meet necessary security requirements. The security framework (for such dynamic environments) must adapt to the different or changing spatial and/or temporal context within the system. This dynamic context information is to be incorporated in the dynamic access model [1, 2]. The authentication context that is used to authenticate the user dynamically helps to identify the user in an active session. The biometric traits used for two-factor authentication [3] introduce traffic in the network. In order to mitigate attacks like man in the middle [4], where an attacker hijacks an active session [5] or gets an access to an unattended personal device to misuse the information, these additional authentication factors [6] are necessary. However, the method of dynamic authentication must reflect the access strategy immediately so that the data is not misused. Such an authorization model that supports the dynamic context change and reflects access decision with respect to the access model is discussed here. In the next section, we discuss the different contexts for authentication, different authorization models for dynamic information adaptation. Then we describe our proposed model with implementation and analysis. We have implemented our proposed model in a healthcare application for remote patients. Patient data is uploaded regularly in the cloud from remote locations (where patients reside) and accessed from other locations regularly/continuously by users like healthcare personnel at health kiosks, doctors, nurses, and patients.

2 Background Details

2.1 Authentication

Authentication is the foremost step for providing security in a cloud environment. However, the user may want to use anonymous authentication strategy that will authenticate him and he can as well avail the service in an unlinkable state. Context like behavioral traits [7] could be used to authenticate the user throughout the access time. This approach will definitely reduce access vulnerability caused by the intrusion. Measurable biometrics like voice, retina, fingerprints, keystroke, etc., distinguish a person. In fact, this kind of authentication can be carried out continuously without even disrupting the user.

In case of continuous authentication scheme, context indicating user behavioral parameters should be handled by a trusted party who is able to validate the details of the user context and subsequently authenticate the user. Biometric information of a regular user is stored in the authentication server. In the active data transfer phase, the captured context is verified with the stored one by the authentication server and the result is made available to the authorization procedure of the system. If any discrepancy occurs, the user is restricted from accessing the resource. Thus, the practice of continuous authentication reduces the risk of access vulnerability. Secure communication between the user and the authentication server is required here for strong security as confidential information is being transferred during the interaction. The implementation contains details about the secure channel setup between user and server.

2.2 Authorization

Authorization is the process of providing access control to a user for resource/s. In the work [8], the researchers classified authorization model into three categories: Static Authorization, Quasi-static Authorization, and Dynamic Authorization. Static authorization model follows the static authentication model, where the user is provided access with a valid authentication credential. Once the user gets access permission, the authorization remains static and ends with an explicit user logout request. This model is, therefore, not suitable for the context-sensitive environment. With advanced security requirement comes the need for session management. The user is granted access for a particular session period, the so-called lease time. Initially, the user is authenticated and authorized for a particular resource access. However, the user has to be validated again, after the lease time expires. The time interval for the lease time is application specific. Dynamic authorization deals with dynamic contextual information. With the change in the context (value), authorization permission changes for the user. However, context information management is the main challenge in this model. A continuous monitoring of the context information has to be carried out and prompt action has to be taken whenever the context (value) changes. This process definitely adds overhead to the access control system. The access control model specifies the privilege of the user over the resource he wants to access. General privileges to access a resource may be stated as permitted or prohibited. However, certain circumstances may allow providing a restricted access or revoking the access based on the present context scenario. Access rules define the action A of a subject S over an object O. The access permission requires a logical relation among the context information. Therefore, generalization of rules for a dynamic environment needs to be incorporated into the system model for simplistic and easy access management. The logical notation of the access permission can be represented as the relation among different context conditions of the subject, object and the action decides the access permission category—permitted, prohibited, or restricted.

2.3 Context-Based Authorization

Static and quasi-static authorization models are not flexible enough with dynamic context change. Hence, adaptability to the dynamic nature of context in the distributed environment should be incorporated. The authors in their work [9] have proposed a dynamic authorization model. In their model, different context observers actively monitor the context behavior and track the changes. When an entity wants to communicate with other entity the context information corresponding to an authorized value is evaluated and access permission is decided.

The permission depends on the prerequisite condition/s and the dynamic values. Access is granted only if all observers are valid and present. They have implemented event-based communication access strategy. However, purely event-based communication generates a huge amount of communication overhead and keeps the server continuously busy with the events. In quasi-static authorization model, the user is provided with an access token for a time period. Within the validity period, the user continues data access, and when the authorization token time expires the user has to generate another access token for data access. Here the user is granted authorization for a lease time. However, in a general quasi-static scenario, the lease time is quite high to provide/revoke access in case any valid context turns invalid. Several security issues crop up in dynamic context information management like the authenticity of the received context information, privacy issues in case of context monitoring, etc. Synchronous communication in context monitoring helps to design the system security.

3 Proposed Model

In the present work, we consider cloud environment and fabricate context-specific access control model using time-triggered [10] methodology. The aim is to make the system less populated with context information with respect to a purely event-triggered [11]-based dynamic authorization model. Therefore, we propose a concept of delta authorization, where the lease time is subdivided into smaller time spans [Lease time $= \delta ti$ for $0<i<n$, where value of n depends on the granularity required for the corresponding context]. With the expiration of each time span the authorization of the user is validated with the dynamic context value. If there is no significant change in the context condition, the authorization is retained. Otherwise the authorization lease time is not extended and the authorization fails with the expiration of the particular time span $\delta ti \ll$ Lease time, for $i>0$.

3.1 Session Data Management in Different Authorization Model

Existence of valid session data is required to receive response data from the cloud provider. Session data keeps the user logged in or out in the system. In Fig. 1, we show the message interaction for security check of the valid session data in different authorization models. For the static authorization model (Fig. 1), once the user has active logon status, he is able to receive the data. Once the user triggers for log out, the data transfer will be stopped. There is not much overhead in managing the session data. In quasi-static authorization model (Fig. 1), with the expiration of the lease time, the session data becomes invalid and data access request stops. However, it is difficult to incorporate the context change within that period. Sometimes it may happen that the context is invalid, but still the user is getting response. In dynamic authorization model (Fig. 2), the active context monitor continuously accesses the context data, validates the data to keep the active session data. But this process will keep the context monitor, the session data manager of the authentication (AUTH) server and authorization (AGS) server for each user, busy all the time. As soon as the user context gets invalid, the user is logged out of the system. This process has huge computational and communication overhead to manage the session data. In our proposed δ Authorization approach (Fig. 2), we consider the lease time (Lt) of the quasi-static authorization model. We divide the lease time into small time spans, δt_i. Each time span is considered to be a small authorization period (-Authorization). Before the beginning of each time span, the context monitor validates the user context and session data is updated based on the result. It implies if the user is authorized for the next delta time span, then the service will be provided to the user. In case the user is not valid, a log out message is triggered to invalidate the session data. And the user is denied response from Cloud Resource Provider for the next time span. Here also we used the concept of buffer data. In case of data transfer from the cloud storage to the user, if the user is authorized for some chunk of data during the delta time span, the data is first buffered, the delta authorization is confirmed, and then data is delivered to the user. Again before transmitting the next chunk of data, next delta authorization is to be confirmed and then the data is delivered to the user/consumer. The use of buffer data helps maintain security in terms of information leakage. Let us consider a doctor asking for some patient health-related data like body temperature, pressure, and ECG value of the next 12 h for each alternate 15-min interval. In the application, the two-factor authentication is used for doctor authentication (i) user credential and (ii) doctor's biometric trait like keystroke validation. Hence, if instead of that doctor any other person pretending to be that doctor tries to read data and provide prescription, the authentication validation of the user (pretending doctor) will fail and data transfer operation will be barred for the session.

3.2   *Structuring the Different Context Conditions*

For further granularity, the context [12] for each entity is subjected to evaluation. This is represented as a function of Con_sub (S), Con_obj (O), Con_action (A). The logical representation defined as:

$$\phi\{\phi(S) \wedge \phi(O) \wedge \phi(A)\} \wedge constraint\_condition(S, O, A)$$

[Here Φ defines the access decision function].

The above relation implies that access permission depends on condition of each entity like context condition of subject, context condition of object, condition of action, and their constraint relationship. This constraint relationship combines all different factors about different contexts of entities and defines their relationship through the access control rule. For a particular access strategy, this constraint needs to be satisfied. Different constraint conditions are designed in multiple ways to provide access control strategy. Co-related constraint can work in a conjunctive fashion, whereas non-correlated contexts can work in a disjunctive fashion. For example, different correlated context conditions of the contexts C1, C2, C3 may be logically applied as

$$OR[\&(C1,C2), \&(C3,C2), \&(C2,C3)] \rightarrow \exists P(Permitted, Restricted, Prohibited)$$

In case of non-correlated context condition the restricted rule could be

$$[AND(C1,C2,C3)] \rightarrow \exists P(Permitted, Restricted, Prohibited)$$

Sometimes it is easier to define the negative rule to define access strategy. In that case negative rule could be defined as

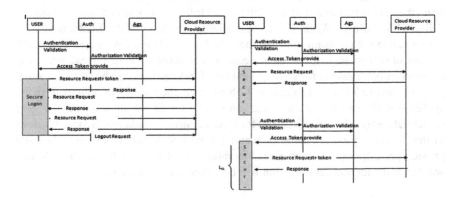

Fig. 1 Message interaction in static and quasi-static authorization

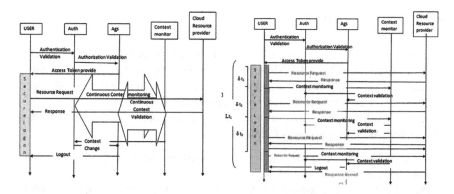

Fig. 2 Message interaction in dynamic authorization and authorization

$$[NOR(C1,C2,C3)] \rightarrow \exists P(Permitted, Restricted, Prohibited)$$

The combination of the correlated or non-correlated condition with their constraint condition rule defines the access permission (permitted, restricted, prohibited) for the access request.

4 Implementation Details

The application considered here is that of health care for remote patients using cloud-based patient data storage and retrieval by relevant users. For a remote healthcare application, a prototype model of the access strategy is implemented. In the application scenario, the patients reside in a remote place. They come to their nearby health center for treatment where their specific health problem, symptoms are captured by health workers. Patient data is uploaded to the cloud storage in a timely manner. The doctor, who is situated remotely from the patient, downloads the patient details and uploads corresponding advice, prescription in the cloud storage server. The health workers download the prescription and thus the treatment is carried out. In this remote healthcare application scenario, a flexible, dynamic access strategy for execution of the entire process is a must to ensure a proper treatment. Until and unless patient data is correctly uploaded and the authenticated doctor provides the treatment, the very purpose of the service is defeated. We need to make sure that, it is only the assigned doctor who is writing the prescription, it is the corresponding particular patient whose data is being checked by the doctor, it is the health worker who is updating the patient health condition, and so on. Otherwise, any stale or improper data could result in a completely wrong treatment. The functional details to implement the security in this scenario for the access control are discussed here with some analysis of the implementation.

4.1   Different Phases

The access control system mainly has these following phases. In our previous work [13], we have introduced the AIC-based anonymous authentication. Some implementation details overlap here with the proposed design. But for the overall system interaction, the details need to be provided for better understanding.

Login: The login phase involves both the authentication server and the Authorization server (AGS). It unfolds as follows:

- At the authentication server, the server validates the user by retrieving their password hash from the user database.
- Once the user is successfully validated, the server generates a unique session ID for the user (provided the user has no session already in progress) and from now, on the user is identified by this session. The user ID session ID mapping is stored locally at the authentication server.
- If the user is validated, then the server generates a fresh bind key for the user.
- The AIC is sent to the AGS (secured using transport security); the AGS responds with the validity period for new AICs. The session is now valid for the duration specified by the AGS.
- A response for the user is generated for the user containing: (a) Confirmation of authentication (b) The authentication token.

Key Access: The user requests keys to access database attributes from the Authorization Grant Server. The AGS interacts with the Key server to serve keys to the user.

- The AGS checks periodically (for each TTL of delta authorization interval) for the existence of the users AIC in its in-memory database; if it exists, check if the AIC has expired or not
- If the AIC exists and is valid then check if the set of access policies allows the user to fulfill the request or not
- If the user is allowed, then validate the user's context [here, we consider user's keystroke information]. The supplied context as part of the key request is decoded and the context validation is carried out.

At the Authentication server, use the context validation sub-module to validate the context (This is not part of the security system) and send back the response.

Back at the AGS, if the Authentication server response was negative, then delete the corresponding AIC from the in-memory database (implicit logout). Else, go forward with key retrieval.

At the Key Server

- Back at the AGS, after successfully getting the key from the Key Server we build a Key Payload for the user
- The users binding key from the AIC is used to encrypt the key and key validity period (as decided by policies at the AGS)

- The users AIC expiration period is extended by the lifetime of the key (This signifies that the user should stay authorized for as long the key is valid). The Key Payload is sent to the user (secured using transport security)

Session Lifetime:

- Here we must make a note about the lifetime of a session. A session is represented as a stored data structure in the in-memory database at the authentication server (this is separate from the in-memory database used to store the AIC at the AGS). The validity of a session is merely the existence of this data structure in the in-memory database. We use an in-memory cache with auto-expiry to invalidate the session data after a fixed period of time. So when a session data structure is deleted after running out of its TTL, the user is effectively logged out. This happens at both the Authentication server and the AGS as both use the same mechanism.
- It is important that such session data be valid for as little time as possible. To do this, the TTL for the session data structure should be as little as possible. After login, the session data is only valid for a very small period of time. To make session management more efficient, this same session data structure is used by extending the lifetime of the data structure.

Data Access:

- When the user requests access to a database attribute, the CSP should check if the user is allowed to do so via the AGS
- CSP determines the access parameters (database, attribute, read/write) and sends them to the AGS along with the user's session ID
- At the AGS, for each TTL of delta authorization interval, the ID is checked against currently authorized users, if not present then the user is not authorized and check if the AIC of the user is still valid or not
- If authorized the CSP allows the operation to go through, else not. The IAM system is CSP agnostic; all we need from the CSP is to do the checking before granting access. This allows the system to be adapted to a variety of different applications without much modification.

Logout: For any dynamic authorization system, we must have an explicit logout method. This is implemented at the authentication server. We use the session key (generated during login) to verify the user session. The need only send a PUT request containing the session key to the Auth server. This prompts the server to validate the session (using the session key), and, if successful, delete the session data structure.

Now this does not delete the AIC stored at the AGS. This is to make sure any delayed data requests still go through even if the user isn't logged in. No new key requests will be served though since the context validation phase will fail (due to the lack of session data in the authentication server). The AIC will not be extended anymore and will invalidate itself later. This is also a reason why the key lifetime should be small enough to not allow this feature to be misused.

5   Comparison

The overall performance of the quasi-static, dynamic, and delta authorization model
depends on a number of issues (Fig. 3).

5.1   Time for Reaction

The difference of our approach with the dynamic approach is the nature of the reac-
tion. The dynamic environment reacts to the event change by proactively checking
for session information before sending the request to the service provider. As shown
in Fig. 3, with increased delta time the execution time gets lower. Here, the task
completion time of different tasks (A, B, C, D, E, F, G) with different delta times has
been shown. However, too high delta time degrades the performance as it enhanced
the risk of access vulnerability. In highly dynamic environment, our approach may
have some delay to reflect the session change information to the service provider,
However, it will not leak the information to an unauthorized user. The proactive
session validation technique and use of a buffer to send data make the system more
secure and flexible.

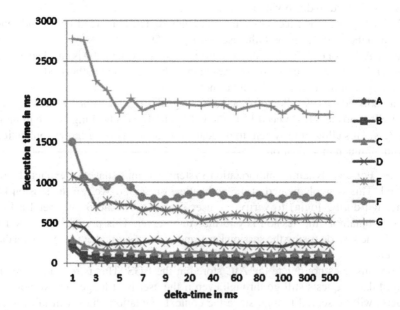

Fig. 3 Performance with varying delta times

Task	Avg Access Time in Proposed Access Control Scheme (ms)	Avg Access Time in Dynamic Access Control Scheme (ms)
A	23	150
B	45	90
C	146	322
D	450	812
E	580	1050

Fig. 4 Performance of dynamic access control and proposed

5.2 Message Overhead

As our approach periodically checks for the session validation instead of continuous validation, the message overhead is much less here with respect to dynamic authorization model. As per our implementation, the message overhead is increased by 16% to 80% for dynamic authorization depending upon the task type shown in Fig. 4. The tasks are categorize based on the execution time. Therefore, the longer the execution time, higher the message overhead. Continuous validation requires a session monitor for each user as service provider side—which may create access vulnerability in the distributed generally untrused environment.

5.3 Session Management

In our approach the session data is managed at the identity and access management server. The AGS validates the context information and modifies the session data. This change is reflected in the service provider side. If the service provider requires to monitor the session data, it requires to keep track for the session of each user—which is not feasible for a flexible, scalable and secure system.

6 Conclusion

Our approach may sometimes produce some delay in reflecting the session information change and invalidating the user, but it would be lesser than the delay introduced in static and quasi-static models. From the perspective of secure data management among different services, our approach provides a better solution since it does not generate huge overhead like dynamic authorization model. However, selection of delta time is very crucial in respect to better throughput. Delta time could be calculated based on different aspects of the task execution. In this paper, we have studied different behaviour of different delta time on different tasks. The authentication context like biometric traits allows users to be validated in an active session. Therefore, hijacking of the active session and acquiring unauthorized service is not allowed in the delta access model. Therefore, the approach will protect from different hazards caused by man in the middle attack (session hijack, get access of the unattended device, etc.).

Acknowledgements This publication is an outcome of the R&D work undertaken in the ITRA project of Media Lab Asia entitled "Remote Health: A Framework for Healthcare Services using Mobile and Sensor-Cloud Technologies".

References

1. Dos Santos, D.R., et al.: Risk-based dynamic access control for a highly scalable cloud federation. In: Proceedings of the Seventh International Conference on Emerging Security Information, Systems and Technologies, SECURWARE (2013)
2. Maa, S., et al.: A trust-based dynamic access control model (2010)
3. Dmitrienko, A., et al.: On the (in) security of mobile two-factor authentication. Financial Cryptography and Data Security, pp. 365–383. Springer, Berlin, Heidelberg (2014)
4. Desmedt, Y.: Man-in-the-middle attack. Encyclopedia of cryptography and security, pp. 759–759. Springer, US (2011)
5. Demchenko, Y., et al.: Web services and grid security vulnerabilities and threats analysis and model. In: Proceedings of the 6th IEEE/ACM international workshop on grid computing. IEEE Computer Society (2005)
6. Kim, J., Hong, S.-P.: A method of risk assessment for multi-factor authentication. J. Inf. Process. Syst. **7**(1), 187–198 (2011)
7. Kathrine, G.J.W., Kirubakaran, E.: Biometric authentication and authorization system for grid security. Int. J. Hybrid Inform. Technol. **4**(4), 43–58 (2011)
8. Tigli, J.-Y., et al.: Context-aware authorization in highly dynamic environments. arXiv preprint arXiv:1102.5194 (2011)
9. Ullah, S., Xuefeng, Z., Feng, Z.: TCloud: a dynamic framework and policies for access control across multiple domains in cloud computing. arXiv preprint arXiv:1305.2865 (2013)
10. Sprinkle, J., Eames, B.: Time-triggered buffers for event-based middleware systems. In: Elissa, K. (ed.) Innovat. Syst. Softw. Eng. **7**(1), 9–22. Unpublished (2011)
11. Albert, A.: Comparison of event-triggered and time-triggered concepts with regard to distributed control systems. Embed. World **2004**, 235–252 (2004)
12. Cuppens, F., Mige, A.: Modelling contexts in the Or-BAC model. Computer Security Applications Conference, 2003. In: Proceedings 19th Annual IEEE, I.S. (2003)

Part III
Wireless Sensor Networks

Two-Hop-Based Geographic Opportunistic Routing in WSNs

Venkatesh, A. L. Akshay, P. Kushal, K. R. Venugopal, L. M. Patnaik and S. S. Iyengar

Abstract Existing work Geographic opportunistic routing (GOR) selects a forwarding sensor node to progress data packets on the basis of geographic distance. Similarly, the multipath routing uses multiple paths to achieve both reliability and delay. However, geographic opportunistic routing results in lower packet delivery rate and high latency. The multipath routing introduces channel contention, interference, and quick depletion of energy of the sensor node in an asymmetric link wireless environment. The existing work *Efficient QoS-aware Geographic Opportunistic Routing (EQGOR)* elects and prioritize the forwarding nodes to achieve different QoS parameters. However, in *EQGOR*, the count of forwarding nodes increases with the increase in the required reliability. To improve energy efficiency, delay, and successful ratio of packet delivery in WSNs, we propose a *Two-Hop Geographic Opportunistic Routing (THGOR)* protocol that selects a subset of 2-hop neighbors of node which has high packet reception ratio and residual energy at the next forwarder node, and the selected 1-hop neighbors of node has supreme coverage of 2-hop neighbors as relay nodes. *THGOR* is comprehensively evaluated through ns-2 simulator and compared with existing protocols *EQGOR* and *GOR*. Simulation results show that *THGOR* significant improvement in packet advancement, delay, reliable transmission, and energy efficient.

Keywords Two-hop packet progress · Geographic opportunistic routing
Media delay · Packet reception ratio

Venkatesh (✉) · A. L. Akshay · P. Kushal · K. R. Venugopal
Department of Computer Science and Engineering, University Visvesvaraya College
of Engineering, Bangalore University, Bengaluru 560001, India
e-mail: venkateshm.uvce@bub.ernet.in; maskija@yahoo.com

A. L. Akshay
e-mail: akshayachar7@gmail.com

P. Kushal
e-mail: kushal.kp80@gmail.com

L. M. Patnaik
National Institute of Advanced Studies, Bengaluru 560012, India

S. S. Iyengar
Florida International University, Miami, FL, USA

© Springer Nature Singapore Pte Ltd. 2018   89
R. Chaki et al. (eds.), *Advanced Computing and Systems
for Security*, Advances in Intelligent Systems and Computing 667,
https://doi.org/10.1007/978-981-10-8183-5_6

1 Introduction

Wireless Sensor Networks (WSNs) is collection of geographically dispersed autonomous sensor nodes which have limited computation and sensing capabilities. Interestingly, there are vast heterogeneity of WSN applications, specifically, environment or terrain observation, war terrain, smart home automation, etc. To ensure reliable transfer and timely communication of data packets from resource-bounded sensor devices to control unit, i.e., sink is a major challenging task in WSNs.

One such challenge is unreliable link of WSN: In real environments, because of interference, attenuation, and channel fading of the unreliable links the traditional routing approaches are not suitable for WSNs. In multipath routing, the data packets are usually copied multiple times and sent to the network. And these packets interfere with each other that reduces the bandwidth, incur congestion at the forwarding nodes. The wireless sensor networks have higher error rate and lower bandwidth than the optical networks. For the recurrent environment describing application, it is a difficult task to successfully deliver the packets on time. Timely and reliable transmission of sensory data is necessary in target tracking and emergency alarm applications. Further, the destination node also expects successful data transmission to be reliable and energy efficient. To accomplish timely and reliable transmission, an accurate and timely update of path quality and routing information are essential. In MMSPEED [1] and MCMP [2], routing algorithms utilize multiple routes between the source and sink pairs. The disjoint multiple routes concept is used to enhance packet delivery in a reliable manner where the end-to-end delay obligation is satisfied as long as an instance of packet reaches the sink within the time limit. Though multipath routing approach provides latency and reliability requirements, it has following two disadvantages: First, *RREQ* route request packets are broadcast to the entire network that leads to high communication overhead, channel contentions increases packet end-to-end delay and depletes sensor node energy quickly. Second, redundancy of data packet on multiple paths though it achieve required reliability but induces significant energy cost, collisions of packets, and congestion in the networks [3].

Motivation: Industrial Wireless Sensor Networks (IWSN) application expect routing protocol to achieve an evenness between energy efficiency, data packet delivery delay and reliable transmission of packet. However, the sensor nodes have limited memory and processing capability. It is essential to develop routing algorithms that have minimum time complexity in potential forwarder set construction and prioritization of forwarding nodes. The existing research routing protocol transmits data over multiple paths to achieve multiobjective [1]. However, the method adopted in these protocols to forward data turns out to be of high-energy consumption. Second, multiple paths result in contention among channels and also introduces interference that increases in delay as well as packet collision [3]. Cheng et al. [4] determine single-hop packet forwarding nodes based on its knowledge of available one-hop neighbor nodes, latency, computation complexity, and energy constraints.

Contribution: Two-Hop packet Geographic Opportunistic Routing (THGOR) provides Expected Packet Progress (EPA) metrics for the selection of the forwarding nodes. The basic idea of selecting a forwarding node is to determine a subset of two-hop neighbors of sender that has expected packet advancement, high probability of successful delivery, and high residual energy. It also select a subset of one-hop neighbor that has ability to cover the selected two-hop forwarding node. THGOR demonstrates the use of optimal sum of forwarding sensor nodes, minimum overhead of control and data packets. With low packet replication overhead, THGOR achieves required reliability, low energy consumption, and end-to-end delay in an efficient way.

Organization: The chapter is organized as follows: A overview of relevant research is discussed in Sect. 2. Background work is explained in Sect. 3. The problem definition and Mathematical model is presented in Sect. 4. Two-Hop geographic opportunistic routing is explained in Sect. 5. Simulation parameters and performance analysis are discussed in Sects. 6 and 7 respectively. Section 8 contains the conclusions.

2 Literature Survey

Data packet routing is a difficult task due to several resource constraints in WSNs. There are several types of routing techniques in WSNs: (i) Hierarchical or tree-based routing, (ii) Heuristic routing and shortest path concepts, (iii) Geographic routing based on node position, and (iv) Operation-based routing. In the tree-based routing, the routing tree is constructed based on QoS parameters and the packets are routed along vertices of tree. However, when the two nodes are in mutual transmission range which belong to different branches that cannot communicate with each other. The sensor nodes near the root node have more energy depletion than others. Therefore, tree-based routing techniques are not energy efficient, even though it is simple and easy to implement.

QoS-aware optimal path is determined using heuristic approach based on shortest path principle. By extended Dijkstra algorithm, least-cost paths are determined which satisfies timeliness and energy requirements. Routing algorithms based on this approach have contention-based scheduling, variable-duty cycle, and traffic-adoptable energy dissipation. However, packet collision overhead leads to retransmission and low packet delivery ratio. Further, enormous number of routing algorithms have been designed in [5–8]. The Geographic Routing is the most encouraging approach for WSNs. Location of the sensor node is utilized to transmit data packets from the source sensor device to the destination or sink [9]. Sensor nodes use immediate neighbors location information to determine the potential forwarder which forwards data packets to the sink node [10–12]. The location details of sensor device and distance among neighboring sensor nodes are determined by received signal strength or GPS of nodes in the network.

In Geographical Adaptive Fidelity (GAF) [12] algorithm, the nodes deployed area is divided into tiny virtual grids. In each virtual grid, all nodes are ranked based on their residual energy. A node with high rank is chosen as an active node, while other nodes turn off their radio. The active sensor node forwards the data packets. Similar to GAF algorithm, Geographic and Energy-Aware Routing (GEAR) algorithm [13] publicize its query directly to a target region through hop-by-hop from the sink node. Each sensor node updates its residual energy and geographical distance to the sink while forwarding the query packets towards the target region. However, recursive geographic query packet forwarding technique may reach dead end or loop forever.

GEographic DIstance Routing (GEDIR) algorithm [14] and Compass Routing [15] use greedy approach to determine the path and ensure that sink receive packets. The QoS provisioning in WSN guarantees that routing layer satisfies various applications requirement like latency, reliability, availability, and security. There are several QoS-aware routing protocols that achieve certain levels of reliability, energy efficiency, and delay requirements by multiple routes among the source and sink node [1, 2, 16, 17]. To maintain data packets confidentiality, data packets over multipath are encrypted using a digital signature crypt system [18–20]. Although, multipath routing reduces routing table updates and enhances packet delivery rate, it results in channel contention and interference [3].

To overcome the limitations of multipath routing, there are several geographic opportunistic routing protocols that shows network performance improvement. In geographic opportunistic routing, any sensor node that overhears the transmission can participate in forwarding the data packets. A set of forwarding nodes at the network layer and one relay node at MAC layer improves the network reliability [21–24]. A set of forwarding nodes are available to forward data packets, but only one forwarding node is selected to forward the data packets; choosing one among them is based on one closest to the sink or one having higher residual energy [25, 26]. Energy efficient and QoS-aware Multipath Routing protocol(EQSR) [27] selects the next forwarding node on basis of the sensor node energy available, available buffer space, and Signal-to-Noise Ratio(SNR). Similarly, forwarding nodes are prioritized based on one-hop progress and reliability [4]. In [28], forwarding nodes are chosen on basis of angle of inclination and distance. Energy-Efficient QoS Assurance Routing (EEQAR) [29] constructs cluster head among the forwarding nodes and achieve evenness in energy utilization by cluster head rotation. WSNs are more susceptible to various attacks because of its broadcast nature and has high error rate than optical communication [30, 31]. Therefore, routing protocol in [32] achieves evenness in energy consumption, and ensures secure data packet delivery. Pratap et al. [33] have analyzed different WSN applications in terms of their significant QOS requirements while Manjula et al. [34] have analyzed different mobility models and its impact on routing algorithm. It is known that a large number of routing algorithms are designed using information on one-hop neighbor. However, every sensor node can have two-hop neighbor information through its one-hop neighbor nodes. The two-hop information based routing algorithms have minimum hop count between the source and sink, the minimum deadline miss ratio, and optimal latency [35]. However, two-hop information-based routing protocols have control packets overhead and

Table 1 Comparison of related works

Authors	Concept	Performance and Advantage	Disadvantage
Niu et al. [38]	It uses bias backoff scheme while route discovery to find the virtual paths	Delivery ratio is improved with high energy efficiency and increases resilience to dynamic links	It computes virtual paths to progress data packets
Altisen et al. [19]	It uses lightweight cryptographic primitives to secure data packets	It improves delivery rate and generates lengthy routes than geographic opportunistic protocol, it is resilient to various attacks	Protocol does not work for dynamic environment with mobile nodes
Gaurav et al. [18]	It distributes traffic among multiple paths, it finds secured disjoint paths using digital signature	Improves delivery delay, transmission rate. It ensures correctness of data at destination	Protocol does not transmit multimedia data, link reliability is not considered
Akkaya et al. [39]	It employs queue and classifies real and non-real time. It associates cost with link.	It increases success rate, reduces delay and energy. Ability to find Qos path for real-time data with delay requirement	It does not consider transmission delay in determining End-to-end delay.
Mao et al. [23]	It selects and prioritize forwarding nodes based on the minimum energy consumption.	It reduces packet duplication ratio and transmission delay. It finds average and maximum delay for each pair node and has less packet loss.	More overhead in sensor nodes in selecting forwarding nodes and does not deliver data most reliably
Proposed THGOR	Based on two-hop packet progress it finds forwarding nodes on the routing path between the source node and destination	Very high success rate Achieves minimum transmission delay, reliable transmission consumes less energy	

high computing complexity for obtaining two-hop neighborhood information [36]. The comparison of related works is given in Table 1 and Bird's eye view of different geographic routing algorithms in WSN is given in Table 2.

The proposed routing protocol *THGOR* obtains information of two-hop neighborhood in circumference of the forwarding area, therefore, proposed routing protocol has average control overhead and computing complexity.

Table 2 Bird's eye view of different geographic routing algorithms in WSN

Year	Authors	Energy Efficiency	Mobility	End-to-end delay	Reliability	Algorithm Complexity
2013	Arafeh et al. [40]	High	No	High	Low	Moderate
2013	Can et al. [41]	High	No	High	Low	Moderate
2014	Zayani et al. [42]	Moderate	No	High	Low	Moderate
2015	Xiuwen et al. [43]	Low	No	Low	Low	Moderate
2015	Khan et al. [44]	High	No	Moderate	Low	High
2015	Cong et al. [45]	Moderate	No	High	Low	High
2015	Sharma et al. [46]	High	No	Low	Low	Moderate
2015	Fucal et al. [47]	High	No	Low	Low	High
2015	Liu et al. [48]	High	No	Low	Moderate	Moderate
2015	Gupta et al. [49]	High	Yes	Low	Moderate	Moderate

3   Background

Geographic Random Forwarding (GeRaF) [25] assigns rank among forwarding candidates based on the single-hop packet advancement. Efficient QoS-aware Geographic Opportunistic Routing (EQGOR) [4] also selects forwarding candidates on the basis of packet reception ratio (PRR), single-hop packet advancement and communication delay. Forwarding candidates residual energy is not considered, while selecting one among the candidate nodes. Anas et al. [37] have evaluated the benefit of opportunistic routing in the presence of unreliable link, loss of DATA, and ACK packets. However, these works address geographic routing with one-hop packet progress toward destination for the multi-constrained WSNs.

4   System Model and Problem Definition

4.1   Definitions of Node's Neighbor and Its Relationships

Sensor nodes form a sensor network $G(SN, L)$ through self-organization, where SN denotes a group of sensors devices and L represents a collection of wireless links. The relationships of sensors nodes are categorized as follows: (i) Inward–outward-neighbor nodes, (ii) Inward-neighbor, (iii) Outward-neighbor, and (iv) outsider.

(1) Let A, B be sensor nodes; if B is in data transmission range of A, and A is in data transmission range of B, then there is direct communication among A to B and B to A; A and B are classified as inward–outward-neighbors and denoted as A ↔ B.

(2) If B is in the data transmission range of A then there is direct communication among A to B; hence, B is inward-neighbor of A and denoted as A → B.
(3) If A is not in the data transmission range of B then there is no communication between B to A, hence, A is outward-neighbor and denoted as A ↤ B.
(4) If there is no communication between A and B, then it is denoted as A ↮ B.

4.2 Terminology: One-Hop and Two-Hop Receivers

Sensor device A's one-hop neighbors are the A's inward-neighbor or inward–outward-neighbor, and one-hop neighbors are in the transmission range r_1. Sensor device A's two-hop receiver are one-hop neighbor of A's one-hop neighbors where two-hop neighbors are in transmission range r_2. Each node determines its neighbor nodes by exchanging Hello messages. Two-hop neighbors information were used in [25, 26] for determining routing path.

4.3 Definition of Forwarding Area

A packet advances from one forwarding node to another forwarding node. Thus, each forwarding node has a transmission range which is denoted as a circle around forwarding node, In the given model, there are two typical forwarding areas. (i) Communication Area (CA), (ii) Degree Radian Area (DRA).

Definition: Communication Area: the range of data transmission for a sensor node is the region where any pair of sensor nodes can hear each other transmissions. r_2: maximum range of transmission for a sensor node. r_1: is minimum transmission range of node i and minimum distance among node i and one-hop neighbor node j.

Definition: Degree Region Area: It is a Θ degree spreading area on both sides of line connecting the sender and sink. The area between two dotted lines is DRA as shown in Fig. 1.

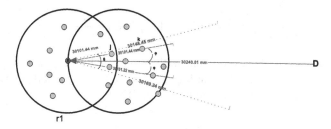

Fig. 1 Node's Degree Region Area(DRA)

5 Mathematical Model

Sensor node distribution is modeled as a spatial poisson process with a constant mean and variance of λ nodes per m^2. Thus, the probability that k nodes present in the area of Am^2 is denoted as

$$Pr(k) = \frac{(\lambda A)^k e^{-\lambda A}}{k!}, \tag{1}$$

where λ is anticipated number nodes in an area.

In *THGOR* protocol, any two nodes in a communication area (CA) are able to over-hear each other's transmissions. The Degree Radian Area (DRA) packet progressing area is $\frac{\pi r^2}{6}$. Thus, Probability of k nodes in DRA is defined as

$$Pr(k) = \frac{(\lambda \frac{\pi r^2}{6})^k e^{-\frac{\pi r^2}{6}}}{k!}, \tag{2}$$

Thus, the average number of one-hop neighbor nodes for a node i (s.t $i \in N_1$) within DRA is ρ: $\rho = \frac{\pi r^2 \lambda}{6}$.

All nodes j belongs ρ are called as Potential Nodes (*PN*). The inclination angle ϕ is determined for each node of *PN*. The inclination angle ϕ is the angle between line connecting node i to $sink(D)$ and line connecting node i to node j. If node j inclination angle ϕ does not exceed the degree $\frac{\Theta}{2}$ and node j is in $r1$, $r2$ then j is identified as Candidate Node (CN$_i$). The inclination angle is determined as: ($Dist_{i,D}$, $Dist_{i,j}$) = | $Dist_{i,D}$ |. |$Dist_{i,j}$)|. $\cos \phi$

$$\phi = \left[\frac{(Dist_{i,D}, Dist_{i,j})}{\| Dist_{i,D} \| . \| Dist_{i,j} \|} \right] \tag{3}$$

When node i decides to transmit data packets to the sink node D, the node i selects j based on the reliability of link. The reliability of link among node i and j is the packet reception ratio among node i and j. This packet reception ratio is calculated on basis of Window Mean Exponential Weighted Moving Average (WMEWMA) [50]. The *prr* is calculated as follows:

$$prr_{i,j} = \beta \times prr_{i,j} + (1 - \beta) \times + \frac{rc}{rc + m}, \tag{4}$$

where rc is the received packets count, m the group of all lost packets and $\beta = 0.6$. The node j is called One-Hop Forwarding Nodes (FN$_1$) if link reliability among node i and node j is more than the threshold point (0.5).

$$RL_{i,j} = prr_{i,j} \tag{5}$$

To ensure that node $j \in$ FN$_1$, it is in region of DRA and it is closer to destination. The distance from node i to node D is greater than transmission range. The following equation is used to ascertain node $j \in FN_1$ is in region of DRA.

$$
\begin{aligned}
Pr(j) &= Pr((dist(i, D) > r) \cap I = 1) \\
&= \int_0^x Pr((dist(i, D) > r) \cap I = 1) d(dist(i, D)) \\
&= \int_r^x Pr(atleastoneFN_1 inDRA) d(dist(i, D)) \\
&= \int_r^x (1 - e^{-\frac{\pi r^2}{6}}) \frac{2 dist(i, D)}{x^2} d(dist(i, D)) \\
&= 1 - \frac{r^2}{x^2} - \frac{2}{x^2} \int_r^x dist(i, D) e^{-\frac{\pi r^2}{6}} d(dist(i, D)),
\end{aligned}
\tag{6}
$$

where I is random variable, and I = 1 if atleast one FN_1 in the DRA and x is the network range. After ascertaining one-hop Forwarding Node j such that $j \in CN_i$ is in DRA. Next, to determine $FN_2(i)$ from two hop neighbors of node i

$$
N_2(i) = \{k : (j, k) \in E \text{ and } j \in FN_1(i),\ k \neq i\}.
\tag{7}
$$

For all one-hop neighbors (say node k) of one-hop Forwarding Node ($FN_1(i)$) that belongs to ρ, the inclination angle ϕ is determined. The inclination angle ϕ is the angle between line connecting node i to $sink(D)$ and line connecting node i to node k. If node k inclination angle ϕ does not exceed the degree $\frac{\Theta}{2}$ and link reliability ($RL_{i,k}$) is greater than two-hop threshold link reliability (0.25) then two-hop neighbor k is included in the candidate nodes set CN_2. The sum of all link reliability of $(FN_1(i), k)$ is computed, where node $k \in CN_2(i)$, resulting sum is multiplied with link reliability between node i and $FN_1(i)$, and the obtained result is subtracted with number of one-hop neighbor of $FN_1(i) \times 0.25$.

$$
TRL_{j \in FN_1(i)} = \{\{prr_{i,j} \times \{\Sigma_{k \in CN_2(i), j \in FN_1(i)} prr_{j,k}\}\} - \mid N(FN_1(i)) \mid \times 0.25\}
\tag{8}
$$

From Eq. (8), link reliability $TRL_{j \in FN_1(i)}$ between node i and its two-hop neighbor nodes nodes belongs to candidate node (CN_2) is determined. Now, potential forwarder is determined by

$$
FN_1(i) = Max\{TRL_j\}.
\tag{9}
$$

In case, two or more nodes of $CN_1(i)$ have same TRL_j then a node is selected from $CN_1(i)$ that cover the maximum number of nodes. Let k nodes be neighbor nodes of $FN_1(i)$ which is denoted as

$$
(k_1, k_2, \ldots, k_n) = \pi N(FN(i)).
$$

A node $FN_1(i)$ selects the next forwarding based on maximum residual energy at node $(N(FN_1(i)))$. The residual energy at the neighboring nodes of $(FN_1(i))$ is calculated based on Eq. (11).

The media delay of each node that belongs to $N(FN(i))$ is derived in Eq. (10). The medium propagation delay is described as time interval from the sender node i broadcasting the packet to the kth ϵ $CN_2(i)$ and forwarding node (kth ϵ $CN_2(i)$) assertion that it has received data packet. This medium propagation delay varies for different MAC protocols and is divided into two parts: (i) Sender delay and (ii) kth ϵ $FN_2(i)$ forwarding node acknowledgement delay. Thus, the medium propagation is given by

$$md = T_c + Tr_d + T_{SIFS} + T_{ack}, \tag{10}$$

where T_c is contention delay, Tr_d is transmission delay, T_{SIFS} is Small InterFrame Space, T_{ack} is acknowledgement delay.

For each forwarding node $k \epsilon CN_2(i)$, the consumption of energy involves the energy utilized to receive and retransmit packets of prior to forwarding sensor nodes to its neighboring node. The available residual energy at two-hop forwarding node $k \epsilon CN_2(i)$ is determined using Eq. (10)

$$E_{k\epsilon N(FN_1(i))}^{re}(t) = E_{k\epsilon N(FN_1(i))}^{i} - ((E_{k\epsilon N(FN_1(i))}^{rc}(t) + E_{k\epsilon N(FN_1(i))}^{tr}(t)) * N_{pkt}) \tag{11}$$

$$FN_2(i) = Max\left(E_{k\epsilon N(FN_1(i))}^{re}(t)\pi\left(N(FN_1(i)) \right) \right). \tag{12}$$

By Eq. (11), the node $FN_2(i)$ is selected as two-hop forwarding node, since this node $FN_2(i)$ satisfies the required reliability and has maximum residual energy. The process of determining next two hops, forwarding sensor nodes is continued iteratively at each two-hop node and routing from the source sensor node to the sink node is through a set of two-hop forwarding nodes.

When the selected two-hop forwarding node fails to deliver packet due to hardware failure then its $FN_1(i)$ selects one of its candidate set as next two-hop forwarding node based on maximum residual energy.

6 Proposed Algorithm

In this section, Two-Hop Geographic Opportunistic Routing(THGOR) is presented. The THGOR determines and prioritizes the two-hop forwarding node using link reliability and optimal energy strategy on each two-hop neighbor of node i, it chooses the optimal one-hop forwarding node as a relay node among candidate nodes $CN_1(i)$.

When node i decides to transmit data packets to the destination node, it identifies its Degree Radian Area (DRA) and one-hop and two-hop neighbor nodes within DRA, and it determines the inclination angle ϕ for each node that belongs to the DRA region. All the nodes having inclination angle less than or equal to $\frac{\Theta}{2}$ and that satisfies link reliability threshold are included in the candidate set $CN_1(i)$ of node i. Further, node i selects and prioritizes the nodes (say $j_1, j_2...j_n$) from the available one-hop neighboring nodes that satisfy link reliability threshold value from node i

to node j_n and it is called as one-hop forwarding node of node i denoted as $FN_1(i)$. The transmitted data packet from node i has a flag bit in its header. The one-hop forwarding node $j \in FN_1(i)$ can distinguish the incoming data packets by tracing the ID of sender and flag bit (line 3), if the flag bit is set to 1 then the received data packets are transmitted to its two-hop forwarding node eventually without adding into the queue and resets the flag bit (line 37).

In the next step, a group of one-hop neighboring nodes of node $FN_1(i)$ and their inclination angle ϕ is determined. The nodes are included into $CN_2(i)$, if $N(FN_1(i))$'s inclination angle is less than or equal to $\frac{\Theta}{2}$. All the nodes that belongs to $CN_2(i)$ and fulfill the threshold value of link reliability and residual energy are ranked (lines 25–30). From $CN_2(i)$, a two-hop forwarding node is selected based on Eq. (12).

The two-hop forwarder node broadcasts its data packet along header setting the flag bit to 1, ID of receiver node, and each two-hop forwarder node in turn iterates the process. On arrival of data packet at one-hop forwarding node, it checks whether a flag bit is set to 1 or 0. If it is set, then the received data packets are transmitted to its two-hop forwarding node eventually without adding into the queue and resets the flag bit. On arrival of data packet at each two-hop forwarder, the routing proceeds and mechanism is repeated to find its two-hop forwarding node.

Estimation of Link Reliability: Function 2 uses exponential weighted moving average based on the window mean [50] to estimate the reliability of link, *RecPkt* is the packets received count, *Pkt.seq* is sequence number of current packet received, f is the packets lost count, *LastPkt* is last packet received, $\frac{RecPkt}{RecPkt+f}$ is the newly determined reliability value. The Rel (N_i, N_j) value is renewed at the receiving node N_j for every size of window.

7 Simulation Setup

To assess the performance of proposed protocol THGOR, the protocol is simulated in NS-2 [51] with C++ code for a different node density. The performance of the proposed protocol is compared with EQGOR [4] and GPSR [11]. The common simulator parameters used during simulations are listed in Table 3. A sink is located at (400, 400 m) and a source sensor node is placed at (0, 0 m). The following performance metrics are used for performance comparison:

- *On time Packet Delivery Ratio*. The ratio of total count of data packets arrived at sink successfully to the total count of packets transmitted by source sensor node.
- *Packet Replication*. Number of redundant packets used to deliver a packet successfully.
- *Two-hop Packet Progress*. Two-hop distance traversed by packets toward destination.
- *Control Packets Overhead*. Number of control packets required in route-discovery process.

Algorithm 1: THGOR:Two-Hop Geographic Opportunistic Routing

Data: E_{avl}, Exp_{rl}, $Rel_{(i,j)}$, $N_1(i)$, $N_2(i)$, flag = 0
Result: Potential Forwarder P_N

1 *Initialization:* $P_N = 0$
2 **while** *(Node! = Sink)* **do**
3   **if** *(flag == 0)* **then**
4        $CN_2(i) = 0$
5        $N_2(i)= N_1(j) - N_1(i)$
6   **for** *($n_k \in N_2(i)$)* **do**
7

$$\phi = \left[\frac{(Dist_{i,D}, Dist_{i,j})}{\| Dist_{i,D} \| . \| Dist_{i,j} \|} \right]$$

8   **if** *($\phi(n_k) \leq \frac{\theta}{2}$)* **then**
9   **if** *($(n_k) \notin (N_1(i), node_j)$)* **then**
10                $CN_2(i) = N_k$
11   **end**
12   **end**
13   **end**
14   Call Link Reliability Estimation:Rel (N_i, N_j)
15   **if** *($Rel_{(i,j)} \geq Exp_{rel}$)* **then**
16   **for** *(j=1 to all one hop Neighbor(i))* **do**
17   *Initialize* $Rel_{Th}[j] =0$
18   **for** *(k=1 to one hop neighbor of $j:N_K$)* **do**
19   **if** *($Rel_{(j,k)} \geq Exp_{rl}$)* **then**
20                   $Rel_{Th}[j] = Rel_{Th}[j] + Rel_{(i,j)} *Rel_{(j,k)}$
21   **end**
22   **end**
23             $Rel_{Th}[j] =Rel_{Th}[j] - (N_2(i) \times Rel_{Th})$
24             $M= Max(Rel_{Th}[j])$
25   **for** *(k=1 to one hop neighbor of M)* **do**
26   Energy required to transmit a packet
27

$$E_{M_k}^{tr} = \left(\frac{e_{M_k}^{el} + e_{M_k}^{tr}}{r} \right)$$

28                $E_{avl}(M,k) = E_{avl}(M) - E_{M_k}^{tr}$
29   **end**
30             $P_N = Max\big(E_{avl}(M,k)\big)$
31   **end**
32          $Node_i$ enters Backoff time
33   **end**
34   set flag
35   **else**
36   act as relay node
37   reset flag
38   **end**
39 **end**

Function 2: Link Reliability Estimation: Rel(N$_i$, N$_j$)

Data: *Node i, Node j, t*
Result: *Pr(Del(N$_i$, N$_j$))*
1 Initialization: *LastPkt = f = RecPkt = 0*
2 **for** *(Each packet(Pkt) arrives at Node$_j$)* **do**
3 | *Increment* RecPkt
4 | f = f + Pkt.sq - (LastPkt+1)
5 | LastPkt = Pkt.sq
6 | Pr(Del(N$_i$,N$_j$)) = $\frac{RecPktint}{t}$
7 **end**

Table 3 Simulation parameters

Parameters	Symbols	Values
Simulation area	sq. meter	400 * 400
Transmission range	r	50 m
TSIFS	μs	10
TDIFS	μs	50
Power required for monitoring events per second	e^{sens}	0.1 mW
Power dissipation to function the wireless	E^{ele}	0.1 mW
The initial available energy at each node	E^{init}	0.05 J
The threshold energy of each node	E^{th}	0.001 J
Packet length	L	1000 bits
Data rate	dr	19.2 kbps
Reliability requirement	r$_{rq}$	0.99
Length of the linear region	D	180 m
End-to-end Delay	T$_{rq}$	0.12 s

8 Performance Analysis

Figure 2 illustrates the effect of number of forwarding nodes on THGOR's performance. It is observed from Fig. 1 that the set of forwarding nodes increases linearly as required reliability increases in GOR and EQGOR [4]. Larger set of forwarding nodes make these protocol more robust since forwarding nodes serve as backup. However, a large set of forwarding nodes result in high percentage of duplicate packets, overhead, and impact of wireless interference. The proposed THGOR protocol yields higher packet delivery ratio with a small set of forwarding nodes with increased reliability. reasons are (*i*) average data transmision link quality, residual energy, and inclination angle of forwarding nodes are taken into account while the selecting forwarding nodes, (*i*) when the forwarding node link reliability is below the threshold value then such node is not considered. The next prioritized forwarding node that is in forwarding area and satisfying the threshold value condition is chosen without backtracking.

Fig. 2 Number of
forwarding nodes *versus*
Reliability Expectation; the
reliability expectation is set
0.99, end-to-end delay is set
to 0.12 s, and range of
transmission for each node is
set to 50 m

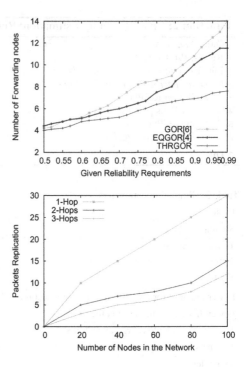

Fig. 3 Average number of
packets replication under
different network nodes,
number of nodes varying
from 20 to 120 with a step
20. Node transmission range
is set to 50 m

Optimal packets replication overhead helps to choose forwarding nodes among the two-hop neighborhood in the route-discovery process is shown in Fig. 3. THGOR has the optimum overhead of packet replication. An optimal packet replication overhead is due to probabilistic strategy to choose forwarding nodes with two-hop neighbor information. The reason is that it uses neighborhood information of every two hops for determining the forwarding node (it does not use one hop information). When the node count varies from 20 to 100, the packet replication overhead for one-hop neighbor information increases since routing decision is made at every one-hop. The packet replication overhead for three-hop information is most stable. However, complexity involved in gathering three-hop information is high. The packet replication overhead with two-hop information is stable with average complexity in gathering information and results in lower end-to-end delay.

Figure 4 illustrates the end-to-end packet successful delivery ratio of EQGOR and THGOR protocol. In the proposed protocol, the packet successful delivery ratio is high. In EQGOR, end-to-end packet delivery ratio is about 70 and 76% when the required reliability is 0.66 and 0.80, respectively, whereas in THGOR, packet delivery ratio is about 73 and 83%. One reason is that the proposed protocol prevents packets from deviating too far toward destination. Another reason is that path length is optimal compared to existing protocol and the routing decision is made at two-hop. EQGOR achieves lower end-to-end packet delivery ratio due to deviation of packets and one-hop routing decision which results in multiplier effect.

Fig. 4 On time packet delivery ratio under different reliability expectation. The reliability Expectation is set 0.99, end-to-end delay is set to 0.12 s, and range transmission for each node is set to 50 m

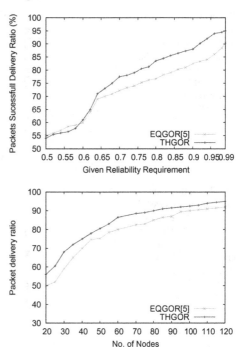

Fig. 5 On time packet successful delivery ratio under a various number of nodes in the network. The number of node varying from 20 to 120, end-to-end delay is set to 0.12 s, and transmission range of each node is set to 50 m

Figure 5 shows that the packet successful delivery ratio grows more or less linearly with the nodes count. The reason is due to the priority assigned among two-hop forwarding nodes based on two-hop packet progress, expected media delay, and residual energy at each forwarding nodes. Compared with the single-hop packet advancement scheme EQGOR, the Two-hop packet progress approach used in THGOR improves the delivery ratio by 6–9% due to selection of two-hop forwarding nodes from Degree Radian forwarding Area(DRA) that are in direction of destination node. Another reason is the probability of void decreases quickly as packet progress at rate of two-hops and probability of collision is less. When the node density is about 60–120, maximum number of nodes are available in forwarding area to become next two-hop forwarding nodes and yields higher packet delivery ratio.

Figure 6 illustrates the packet progress toward intended destination. EQGOR uses single-hop packet progress and achieves progress toward the destination is about 36–31 m when reliability requirement varies from 0.5 to 0.99. For the same reliability, in THGOR, the packet progress toward destination is about 46–44 m, the selected forwarding nodes are selected the most reliable link. Another reason is when forwarding node (*say k*) fails to transmit a packet successfully, a retransmission is initiated from the immediate previous node instead of going back to the source node. As the packet progresses faster towards the destination, the sender locates more of forwarding nodes to forward the packets which result in low end-to-end delay.

It is observed that in Fig. 7, two-hop progress with the two-hop information performs better in terms of reachability with less number of transmissions, since the

Fig. 6 Comparison of
Two-hop packet progress
under different reliability
requirement. The reliability
expectation is set 0.99, delay
is set to 0.12 s, and range of
transmission of each node is
set to 50 m

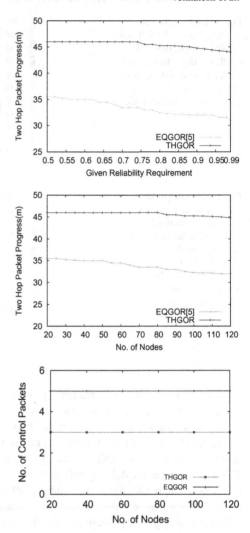

Fig. 7 Comparison of On
time packet successful
delivery ratio under different
node density. The number of
node varying from 20 to 120,
end-to-end delay is set to
0.12 s, and transmission
range of each node is set to
50 m

Fig. 8 Average number of
control packet over under
different node density,
Control packets include
HELLO, RTS,CTS, and
ACK packets

proximity of two-hop neighbor information is considered during routing decisions.
The packet progress significantly increases at the low density (from 20 to 70 nodes).
As the number of nodes increase (from 71 to 120 nodes), packet progress is low,
the void distance between the two nodes decreases, improving the quality of link
between the two nodes. In EQGOR, the routing decision is made at each node and
it neglects the reliability of links, hence the packet progress is comparatively lower
than THGOR wherein proximity of two-hop node information and the link quality
is taken into the account. The two-hop packet advancement from the source sensor
node to destination node is a crucial factor in view of the delay, consumption of
energy and hop count. Figure 8 illustrates the average number of control packets
exchange between the forwarding nodes. The control packets include RTS, CTS,

ACK, and HELLO messages to identify neighbors and its corresponding PRR value. The control packets cost is directly proportional to the number of data transmissions. In EQGOR, the overhead of control packets increases linearly with the number of nodes in the network for the following reasons:

- More nodes are involved in its periodic flooding to determine neighbor node information and a forwarding node sends data packets in random directions,
- The number of updates on neighboring node and link quality in EQGOR is larger than THGOR since only nodes that are in forwarding area have to update its neighboring nodes, inclination angle, and link quality,
- The link reliability update is quite small in THGOR compared to the number of link quality update in EQGOR.

9 Conclusions

Simple geographic opportunistic routing uses its local knowledge of next-hop to forward the packets. However, this may lead to transmission failure due to low link reliability, and more retransmissions. The proposed protocol THGOR uses two-hop reliability and residual energy packet progress. The results of simulation shows there is a clear fast packet progress towards destination and decline in the average number of transmissions with low end-to-end delay while selecting the next forwarding nodes. THGOR strikes balance between packet progress, computation complexity. Extensive simulation results show that THGOR outperforms the EQGOR [4] and GOR [25] protocols.

References

1. Felemban, E., Lee, C.G., Ekici, E.: MMSPEED: Multipath multi-speed protocol for QoS guarantee of reliability and timeliness in wireless sensor networks. IEEE Trans. Mob. Comput. **5**(6), 738–754 (2006)
2. Huang, X., Fang, Y.: MCMP: multi-constrained QoS multipath routing in wireless sensor networks. Wirel. Netw, **14**(4), 465–478 (2008)
3. Wang, Z., Bulut, E., Szymanski, B.K.: Energy efficient collision aware multipath routing for wireless sensor networks. In: IEEE International Conference on Communications-ICC'09, pp. 1–5 (2009)
4. Cheng, L., Niu, J., Cao, J., Das, S.K., Gu, Y.: QoS aware geographic opportunistic routing in wireless sensor networks. IEEE Trans. Parallel Distrib. Syst. **25**(7), 1864–1875 (2014)
5. Tarannum, S., Aravinda, B., Nalini, L., Venugopal, K.R., Patnaik, L.M.: Routing protocol for lifetime maximization of wireless sensor networks. In: Proceedings of IEEE International Conference on Advanced Computing and Communications, pp. 401–406, 20–23 Dec 2006
6. Tarannum, S., Srividya, S, Asha, D.S., Padmini, R., Nalini, L., Venugopal, K.R., Patnaik, L.M.: Dynamic hierarchical communication paradigm for wireless sensor networks: a centralized energy efficient approach. In: Proceeding of 11th IEEE International Conference on Communication System, pp. 959–963, 19–21 Nov 2008

7. Kanavalli, A., Shenoy, P.D., Venugopal, K.R., Patnaik, L.M.: A flat routing protocol in sensor networks. In: Proceedings of International Conference on Methods and Models in Computer Science, pp. 1–5, 14–16 Dec 2009. ISBN: 978-1-4244-5051-0

8. Manjula, S.H., Abhilash, C.N., Shaila, K., Venugopal, K.R., Patnaik, L.M.: Performance of AODV routing protocol using group and entity mobility models in wireless sensor networks. In: Proceedings of the International MultiConference of Engineers and Computer Scientists, vol. 2. Hong Kong, 19–21 Mar 2008

9. Li, Y., Li, J., Ren, J., Wu, J.: Providing hop-by-hop authentication and source privacy in wireless sensor networks. In: Proceedings of IEEE INFOCOM 2012, pp. 3071–3075 (2012)

10. Li, J., Jannotti, J., De Couto, D.S., Karger, D.R., Morris, R.: A scalable location service for geographic ad-hoc routing. In: Proceedings of ACM 6th Annual International Conference on Mobile Computing and Networking, pp. 120–130 (2000)

11. Karp, B., Kung, H.T.: GPSR: greedy perimeter stateless routing for wireless networks. In: Proceedings of ACM 6th Annual International Conference on Mobile Computing and Networking, pp. 243–254 (2000)

12. Xu, Y., Heidemann, J., Estrin, D.: Geography-informed energy conservation for ad-hoc routing. In: Proceedings of ACM 7th Annual International Conference on Mobile Computing and Networking, pp. 70–84. ACM

13. Yu, Y., Govindan, R., Estrin, D.: Geographical and energy-aware routing: a recursive data dissemination protocol for wireless sensor networks. Technical Report ucla/csd-tr-01-0023, UCLA Computer Science Department, May 2001

14. Lin, X., Stojmenovic, I.: Geographic distance routing in ad-hoc wireless networks. IEEE J. Sel. Areas Commun. (1988)

15. Kranakis, E., Singh, H., Urrutia, J.: Compass routing on geometric networks. In: Proceedings of 11th Canadian Conference on Computational Geometry (1999)

16. Razzaque, M.A., Alam, M.M., Or-Rashid, M., Hong, C.S.: Multiconstrained QoS geographic routing for heterogeneous traffic in sensor networks. Proc. CCNC **2008**, 157–162 (2008)

17. Marina, M.K., Das, S.R.: Ad hoc on-demand multipath distance vector routing. In: Proceedings of International Conference on Wireless Communications and Mobile Computing, pp. 969–988 (2006)

18. Gaurav, S.M., D'Souza, R.J., Varaprasad, G.: Digital signature-based secure node disjoint multipath routing protocol for wireless sensor networks. IEEE Sens. J. **12**(10), 2941–2949 (2012)

19. Altisen, K., Devismes, S., Jamet, R., Lafourcade, P.: SR3: secure resilient reputation-based routing. In: Proceedings of IEEE International Conference on Distributed Computing in Sensor Systems (DCOSS2013), 2013, pp. 258–265 (2013)

20. Alrajeh, N.A., Alabed, M.S., Elwahiby, M.S.: Secure ant-based routing protocol for wireless sensor networks. Int. J. Distrib. Sens. Netw. **2013**, Article ID 326295 (2013)

21. Rozner, E., Han, M.K., Qiu, L., Zhang, Y.: Model-driven optimization of opportunistic routing. ACM SIGMETRICS Perform. Eval. Rev. **39**(1), 229–240 (2011)

22. Zeng, K., Lou, W., Yang, J.: On throughput efficiency of geographic opportunistic routing in multihop wireless networks. J. Mob. Netw. Appl. **12**(5), 347–357 (2007)

23. Mao, X., Tang, S., Xu, X., Li, X.Y., Ma, H.: Energy-efficient opportunistic routing in wireless sensor networks. IEEE Trans. Parallel Distrib. Syst. **22**(11), 1934–1942 (2011)

24. Landsiedel, O., Ghadimi, E., Duquennoy, S., Johansson, M.: Low power, low delay: opportunistic routing meets duty cycling. Proc. IPSN **2012**, 185–196 (2012)

25. Zorzi, M., Rao, R.R.: Geographic random forwarding (GeRaF) for ad hoc and sensor networks: energy and latency performance. IEEE Trans. Mob. Comput. **2**(4), 349–365 (2003)

26. Park, J., Kim, Y.N., Byun, J.Y.: A forwarder selection method for the greedy mode operation of a geographic routing protocol in WSN. In: Ubiquitous and Future Networks(ICUFN), 2013, pp. 270–275, July 2013

27. Yahya, B., Ben-Othman, J.: An energy efficient and QoS aware multipath routing protocol for wireless sensor networks. In: IEEE 34th Conference on in Local Computer Networks, pp. 93–100 (2009)

28. Spachos, P., Toumpakaris, D., Hatzinakos, D.: QoS and energy-aware dynamic routing in wireless multimedia sensor networks. In: IEEE International Conference on Communications(ICC)-2015, pp. 6935–6940 (2015)
29. Lin, K., Rodrigues, J.J.P.C., Ge, H., Xiong, N., Liang, X.: Energy efficiency QoS assurance routing in wireless multimedia sensor networks. IEEE J. Syst. **5**(4), 495–505 (2011)
30. Venugopal, K.R., Ezhil Rajan, E., Sreenivasa Kumar, P.: Impact of wavelength converters in wavelength routed all-optical networks. Comput. Commun. **22**(3), 244–257 (1999)
31. Venugopal, K.R., Srinivasa, K.G., Patnaik, L.M.: Soft computing for data mining applications. Springer (2009). ISBN 978-3-642-00192-5, e-ISBN 978-3-642-00193-2
32. Tang, D., Ren, J., Wu, J.: Cost-aware secure routing (CASER) protocol design for wireless sensor networks. IEEE Trans. Parallel Distrib. Syst. **26**(4), 960–973 (2015)
33. Pratap, U., Shenoy, P.D., Venugopal, K.R.: Wireless sensor networks applications and routing protocols: survey and research challenges. In: Proceedings of International Symposium on Cloud and Services Computing, pp. 49–56, Dec 2012
34. Manjula, S.H., Abhilash, C.N., Shaila, K., Venugopal, K.R., Patnaik, L.M.: Performance of AODV routing protocol using group and entity mobility models in wireless sensor networks. In: Proceedings of the International MultiConference of Engineers and Computer Scientists, vol. 2, pp. 1212–1217 (2008)
35. Chen, C.S., Li, Y., Song, Y.Q.: An exploration of geographic routing with k-hop based searching in wireless sensor networks. In: Proceedings of IEEE Conference on Communications and Networking, pp. 376–381 (2008)
36. Jung, J., Park, S., Lee, E., Oh, S., Kim, S.H.: OMLRP: multi-hop information based real-time routing protocol in wireless sensor networks. In: IEEE Wireless Communications and Networking Conference (WCNC), pp. 1–6 (2010)
37. Basalamah, A., Kim, S.M., Geuo, S., He, T., Tobe, Y.: Link correlation aware opportunistic routing. In: IEEE Proceedings INFOCOM, 2012, pp. 3036–3040 (2012)
38. Niu, J., Cheng, L., Gu, Y., Shu, L.: R3E: reliable reactive routing enhancement for wireless sensor networks. IEEE Trans. Ind. Inf. **10**(1), 784–794 (2014)
39. Akkaya, K., Younis, M.: An energy-aware QoS routing protocol for wireless sensor networks. In: The Proceedings of IEEE ICDCSW, pp. 710–715, May 2013
40. Arafeh, B., Day, K., Touzene, A., Alzeidi, N.: GEGR: a grid-based enabled geographic routing in Wireless Sensor Networks. In: IEEE Malaysia International Conference on Communications (MICC), pp. 367–372 (2013)
41. Ma, C., Wang, L., Xu, J., Qin, Z., Shu, L., Wu, D.: An overlapping clustering approach for routing in wireless sensor networks. In: IEEE Conference on Wireless Communications and Networking(WCNC), pp. 4375–4380 (2013)
42. Zayani, M.-H., Aitsaadi, N., Muhlethaler, P.: A new opportunistic routing sheme in low duty-cycle WSNs for monitoring iinfrequent events. In: IEEE Conference on Wireless Days (WD), pp. 1–4 (2014)
43. Fu, X., Li, W., Ming, H., Fortino, G.: A framework for WSN-based opportunistic networks. In: IEEE 19th International Conference on Computer Supported Cooperative Work in Design (CSCWD), pp. 343–348 (2015)
44. Khan, G., Gola, K.K., Ali, W.: Energy efficient routing algorithm for UWSN-A clustering approach. In: IEEE Second International Conference on Advances in Computing and Communication Engineering (ICACCE), pp. 150–155 (2015)
45. Sun, C., Zhu, Y.-H., Yuan, L., Chi, K.: Borrowing address from two-hop neighbor to improve successful probability of joining IEEE 802.15. 5-based mesh wireless sensor networks. In: IEEE 7th International Conference on New Technologies, Mobility and Security (NTMS), pp. 1–7 (2015)
46. Sharma, M., Singh, Y.: Middle position dynamic energy opportunistic routing for wireless sensor networks. In: IEEE International Conference on Advances in Computing, Communications and Informatics (ICACCI), pp. 948–953 (2015)
47. Yu, F., Pan, S., Hu, G.: Hole plastic scheme for geographic routing in wireless sensor networks. In: Proceedings of IEEE International Conference on Communications (ICC), pp. 6444–6449 (2015)

48. Liu, C., Fang, D., Chen, X., Hu, Y., Cui, W., Xu, G., Chen, H.: LSVS: bringing layer slicing and virtual sinks to geographic opportunistic routing in strip WSNs. In: Proceedings of IEEE Fifth International Conference on Big Data and Cloud Computing (BDCloud), pp. 281–286 (2015)
49. Gupta, H.P., Rao, S.V., Yadav, A.K., Dutta, T.: Geographic routing in clustered wireless sensor networks among obstacles. IEEE Sens. J. **5**, 2984–2992 (2015)
50. Woo, A., Culler, D.E.: Evaluation of efficient link reliability estimators for low-power wireless networks. Technical Report. University of California, Computer Science Division, 2003
51. Venugopal, K.R., Rajakumar, B.: Mastering C++. 2nd edn. McGraw Hill Education. pp. 881 (2013). ISBN(13): 978-1-25902994-3, ISBN(10):1-25-902994-8
52. Bulusu, N., Heidemann, J., Estrin, D.: GPS-less low-cost outdoor localization for very small devices. IEEE Trans. Pers. Commun. **7**(5), 28–34 (2000)

Data Gathering from Path-Constrained Mobile Sensors Using Data MULE

Dinesh Dash and Naween Kumar

Abstract In Wireless Sensor Network (WSN), sensor nodes are deployed to sense useful data from environment. Sensors are energy-constrained devices. To prolong the sensor network lifetime, nowadays mobile robots (sometimes referred as data sink, data mules, or data collectors) are used for collecting the sensed data from the sensors. In this environment, sensor nodes may directly transfer their sensed data to the data mules. Sensed data are sometimes time sensitive; therefore, the data should be collected within a predefined period. Hence, depending on the speed of the data mules, the trajectory lengths of the data mules have upper limits. In this paper, an approximation algorithm is proposed for collecting data from the mobile sensors using data mules.

Keywords Mobile sink · Data gathering protocol · Wireless sensor network
Approximation algorithm

1 Introduction

Wireless Sensor Network (WSN) consists of large number of sensors (nodes) and few Base Stations (BSs). Each sensor has a sensing range and a communication range. Within the sensing range, sensor can sense environmental data and it can communicate to other sensors, which are within its communication range. A typical application in WSN is to collect the sensed data from individual sensors to a BS. Depending on the communication range of the sensors, they form a communication network topology. Two nodes are connected by an edge, if they are within their

D. Dash (✉) · N. Kumar
Department of Computer Science & Engineering, National Institute
of Technology Patna, Patna 800005, Bihar, India
e-mail: dd@nitp.ac.in

N. Kumar
e-mail: kr.naween@gmail.com

© Springer Nature Singapore Pte Ltd. 2018
R. Chaki et al. (eds.), *Advanced Computing and Systems
for Security*, Advances in Intelligent Systems and Computing 667,
https://doi.org/10.1007/978-981-10-8183-5_7

communication range. Most of the nodes cannot communicate directly to the BS and they send data packet to BS through multi-hop communication.

Hierarchical or cluster-based routing methods are proposed in wireless networks, in which a subset of nodes is selected to form communication layer topology, and only the nodes in the communication layer participate for data communication and thereby reduce the transmission overhead of redundant information. It simplifies the topology of the network and saves energy for information gathering and forwarding.

Data collection is one of the fundamental operations in WSN. Other critical network operations such as event detection, robust message delivery, localization and network reconfiguration, etc. are depended on data collection as a basic operation. Data aggregation and in-network processing techniques have been investigated recently as efficient approaches to achieve significant energy savings in WSN by combining data arriving from different sensor nodes at some aggregation points, eliminating redundancy, and minimizing the number of transmissions before forwarding data to the sinks. Hence, data fusion or aggregation has emerged as a useful paradigm in sensor networks. Due to the multi-hop data transmission in static sink based WSN, unbalanced energy consumption is caused by the nodes close to sink and other distant sensor nodes. Sensor nodes, which are closed to the sink node, have to carry much more traffic overhead compared with distant sensor nodes. Since sensor nodes are limited to battery power supply, such unbalanced energy consumption causes quick power depletion on part of the network and reduced the lifetime of the network. To resolve this issue, recent research works have been proposing mobile sink based data gathering techniques.

Mobile sink is an unmanned vehicle/robot that roams around the area and collects sensed data from data collectors. Mobile sink based data gathering techniques are also useful in applications involving real-time data traffic. In such applications, data gathering paths by the mobile sink are selected so that certain end-to-end delay constraints are satisfied. In order to improve the round-trip time of the mobile sink, one of the solutions could be to move the sink only to few data collectors rather than individual sensors. To fulfill the time constraint of real-time sensor data, a subset of

Fig. 1 Path-constrained
mobile sensor network

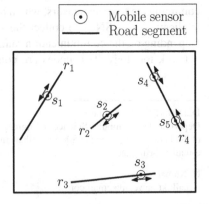

sensors called cluster heads are selected efficiently so that the total length to visit them for collecting the data is minimum.

In path-constrained mobile sensor network, sensors are moving along a set of predefined paths (roads). Mobile sink can move to any arbitrary position for collecting data. An example of path-constrained mobile sensor network is shown in Fig. 1. Paths are denoted by $\{r_1, r_2, ..., r_4\}$ (paths for the mobile sensors), and circles $\{s_1, s_2, ..., s_5\}$ denote mobile sensors. The mobility of the sensors is confined within the segments. Data gathering problem from mobile sensors is a challenging problem.

1.1 Contribution

In this paper, our contributions to mobile data gathering problems are as follows:

- We have addressed periodic data gathering protocol from a set of mobile sensors where trajectory of the mobile sensors is confined within a set of straight-line segments on a plane.
- We have identified two limitations of the solution proposed in [1] and proposed a solution to overcome the limitations.
- A 4-factor approximation algorithm is proposed for the above problem.

The rest of the paper is organized as follows. Section 2 briefly describes related works of the paper. Formal definition of problem and network model is described in Sect. 3. Our solution approach is discussed in Sect. 4. Section 5 concludes the paper and describes some of its future works.

2 Related Works

In wireless sensor networks, data gathering protocols are broadly classified into two types: static sink node based and mobile sink node based. In static sink architecture, all static sensors send the sensed data to the static sink node through multi-hop communication, whereas in mobile sink architecture, the sensed data are collected by a mobile sink node by visiting the communication range of individual sensors after a certain time interval.

In static sink architecture, there are two types of nodes: regular node and sink node. Regular node senses the data and sends the data to the sink node. There is a data aggregation tree rooted at the sink. The tree is generated from the communication topology of the sensor network. In [2], Singh et al. proposed a data collection scheme in a grid-based clustering network with mules to reduce hot spot problem. He et al. [3] proposed a load balanced data aggregation tree for gathering the sensed data from a probabilistic network model.

In mobile sink based data gathering protocols, the mobile sink instead of visiting all static sensors chooses a subset of sensors as gateways. Therefore, in this type of network, there are three types of sensor nodes: regular node, intermediate gateway node, and sink node. Regular nodes sense the environmental data and send it either to a sink or to an intermediate gateway node. Gateway node works as regular node as well as it also helps to forward other sensors data.

In [4], Liu et al. proposed a cluster-based solution for finding optimal path of mobile sink to collect the data from the individual sensors. In this protocol, there is no restriction on path of the mobile sink. The sensors with common intersection on their communication range form a cluster. The mobile sink only visits the common intersection zones to collect the sensed data. It will collect the sensed data directly from each individual sensor. So there is no need of any intermediate gateway node. The total trajectory of the mule is minimum. An improved version of the genetic algorithm is proposed in [5]. In this work, authors proposed a novel method for population generation in the genetic algorithm and thereafter, effective shortcut techniques are proposed to improve the path length.

In [6], Kim et al. proposed approximation algorithm to collect data from static sensors using data mules. Data mules collect data from the sensor within its neighborhood. To reduce the data gathering paths of the data mules, multiple data mules are used. To reduce data gathering period, maximum trajectory length of the data mules is minimized. Gao et al. in [7] presented a data gathering algorithm from a set of static sensor nodes using a mobile data collector. Path of the mobile data collector (sink) is a fixed path P.

The objective of the work is to find a continuous sub-path P' on the path P of length $V \times T/2$ for mobile sink node, where V is velocity of the mobile sink and T is the time deadline of the sensed data, such that the total energy consumption by the network to deliver the data to the sub-sink nodes close to the path P' is minimum. The sub-path based on the maximum number of sub-sink is closed to the path within predefined travel distance. In [8], Mai et al. proposed a load balance mobile sink based data gathering protocol where the mobile sink runs with a given speed and wants to finish its data gathering cycle within a given time bound. They assumed that there is no restriction on the path of the mobile sink. The objective is to select a set of gateways such that loads on the gateways are almost same and the trajectory length of the mobile sink satisfies the maximum delay.

Gao et al. [9] proposed a path-constrained data gathering protocol. They tried to maximize the amount of data collected per round of the mobile sink. As the mobile sink roams around a predefined path with a fixed speed, it will stay for a fixed amount of time close to any intermediate gateway node. Therefore, if the intermediate gateway node has too many data, it will be unable to transfer the whole data to the mobile sink within that time period. In order to increase the total amount of gathered data by the mobile sink, the data sensed by sensors must be uniformly distributed among the gateways. ILP formulation of the problem is shown, and a genetic algorithm is proposed. In [10], a subset of sensors is considered as a data sources, and all of them generate same amount data.

Different sensors have different time deadlines, within the time deadline, data need to be delivered to a predefined sink node. The goal is to find energy-efficient set of paths from the sources to the sink, which will satisfy the time deadline as well as the total energy consumption is minimum.

Data gathering from mobile sensors is another challenging area in the sensor network. In [1], a solution for data gathering from path-constrained mobile sensors is proposed using data mules. The paths of the mobile sensors are restricted along a set of line segments and the mobile sink visits all the mobile sensors to collect their sensed data.

3 Problem Statement and Complexity Analysis

A set of mobile sensors $S = \{s_1, s_2, ..., s_N\}$ are moving along a set of road segments $R = \{r_1, r_2, ..., r_M\}$. Assume a data mule can collect data from a mobile sensor when it reaches to the point where mobile sensor presents. Again, assume that the movement paths of the mobile sensors are arbitrary along the R. The movement speed of the mobile sensors is also arbitrary. Sensors can stop movement for arbitrary time. Data mules can move to any location on plane with a fixed speed V.

Problem 1 Find minimum number of data mules and their traveling paths to collect the sensed data from all mobile sensors within a specified time period t.

Theorem 1 *The Problem 1 is NP-hard unless P = NP.*

Proof Finding minimum number of data mules to collect data from all the mobile sensors moving on a set of segments is NP-hard. If the line segments' lengths are reduced to zero, then the line segments become points and the mobile sensors become static because the path of the mobile sensors is restricted to point. Therefore, data gathering from a set of static sensors and delivering it to BS within time t using minimum data mules is a special case of Problem 1. The decision version of the problem can be stated as follows: Is there a data gathering tour of k mules with time period t (is equivalent to a tour of length of at most $k \times Vt$) which includes all static sensors and the BS. Therefore, the decision version of the special case is an instance of traveling salesman problem which is NP-complete [11]. Hence, Problem 1 is NP-hard.

4 Algorithm to Find Minimum Data Mules and Their Paths

In this section, we discuss a data gathering protocol using mobile data collectors to collect data from a set of mobile sensors which are moving on a set of road segments. We refer the algorithm as Data Gathering from Path-Constrained Mobile Sensors (DGPCMS). Gorain et al. [1] proposed an approximation algorithm for data gathering from path-constrained mobile sensors. But the proposed solution has limitation: the length of the road segments is bounded by $Vt/2$, where V is the speed of the data mules and t is the data gathering time interval. In this paper, we have addressed these limitations and proposed an approximation solution for the problem, which is based on the following characteristic.

Lemma 1 *Every point of all road segments must be visited by a data mule at least once within time interval t.*

Proof Assume that there is a point P on a road segment r_i, which is not visited by any mobile data collector within time interval t and still all mobile sensors are visited by the mobile collectors. A mobile sensor can move to P and stay there for more than t time. Therefore, no mobile collector will able to collect data from the mobile sensor at P. It contradicts our assumption. Hence, the lemma is true.

Determine the shortest distance c_{ij} between end points of every pair of segments (r_i, r_j), where i, j and call them as intersegment connectors. Construct a complete graph among the segments with the help of intersegment connectors. Our algorithm determines a solution of the number of data mules requirement in M rounds, where M is the number road segments. In particular, round k (k varying in between 1 and M) finds a minimum spanning forest F_k with $(M - k + 1)$ trees by interconnecting the segments in R with $(k - 1)$ intersegment connectors. For each tree $T_i \in F_k, i \in \{1 \ldots M - k + 1\}$ construct Euler cycle EC_i by doubling each edge of T_i and then, remove the highest length intersegment connector from EC_i to convert it to Euler path E_i. Let N_k denote an upper bound on the number of data mules requirement by our algorithm to traverse $(M - k + 1)$ Euler paths in kth round. Divide $(M - k + 1)$ Euler paths into segments of equal length of size at most Vt. Thereafter, deploy two data mules at the two ends of every segment and they start moving toward each other until they meet at the middle and thereafter reverse their movement directions until they reach their starting points again. Since the length of Euler path $L(E_i) \leq 2L(T_i)$ for $i \in \{1, 2, \ldots (M - k + 1)\}$, we can write $N_k \leq 2 \sum_{i=1}^{M-k+1} \lceil \frac{2L(T_i)}{Vt} \rceil$, where $L(T_i)$ denotes the length of tree T_i and Vt denotes the distance travel by a data mule within t time period. The detailed algorithm is shown in Algorithm 1.

4.1 Example of Execution of Algorithm DGPCMS

The sensor network in Fig. 1 has four segments; four rounds are required for determining the usage of data mules, as shown in Fig. 2. Initially, in round 1, all the road segments are considered as independent tree as shown in Fig. 2a. Trees $\{T_1, T_2, T_3, T_4\}$ are converted into Euler paths $\{E_1, E_2, E_3, E_4\}$. Determine the number of data mules requirement $N_1 = 2 \sum_{i=1}^{4} \lceil \frac{L(E_i)}{Vt} \rceil \leq 2 \sum_{i=1}^{4} \lceil \frac{2L(T_i)}{Vt} \rceil$, where $L(E_i)$ denotes the length of Euler path E_i and Vt denotes the distance traveled by a data mule in time t. In round 2, reduce the number of trees by interconnecting closest pair of trees $\{T_2, T_4\}$ of round 1 (connected by a dotted intersegment connector) and the new tree is referred as T_2. In this round, there are three trees $\{T_1, T_2, T_3\}$ and their corresponding Euler paths E_1, E_2, E_3 are shown in Fig. 2b. Repeat the same process for round 3 and 4 with two trees and subsequently for one tree, as shown in Fig. 2c, d, respectively. Let J be the index of a round for which number of data mules requirement is minimum ($N_J = \text{Min}\{N_1, N_2, N_3, N_4\}$). According to our algorithm, partition the Euler paths of the Jth round into sub-paths of equal length of size at most Vt. An example of movements of data mules to collect data from mobile sensors from an Euler path is shown in Fig. 3. In Fig. 3a, Euler path is split into three sub-paths by two perpendicular segments. And for every sub-path, two data mules are deployed at the two end points; and they start moving toward each other at the starting time $t' = 0$. At time $t' = t/2$, the two data mules meet with each other at middle of the sub-path and change their movements in opposite directions, as shown in Fig. 3b.

Algorithm 1: DGPCMS(Data Gathering From Path Constrained Mobile Sensors)

Input : Set of road segments $R = \{r_1, r_2, \ldots, r_N\}$

Output : Paths for the mobile sinks

1   for $k \leftarrow 1$ to M do

2   Find a minimum spanning forest F_k by interconnecting the end points of segments in

N R with $(k-1)$ inter segments connectors. Let $\{T_1, T_2, \ldots, T_{(M-k+1)}\}$ be the tree components of F_k.

3 $k = 0$

/* Number of data mules used to collect data from k Euler paths. */

4   for $i \leftarrow 1$ to $(M-k+1)$ do

5 $ST_i = \left\lceil \frac{(2L_{(T_i)})}{Vt} \right\rceil$ /* Number of sub paths for kth Euler path

*/

6 end

7 $N_k = N_k + 2 \times ST_i$

8 end

9 Let J be the index in between 1 to M such that $N_J = MIN\{N_1, N_2, \ldots, N_M\}$.

10 Construct Euler path E_i for each tree $T_i \in T_J$ by making double each edge of T_i except the maximum length edge in T_i.

11   for $i \leftarrow 1$ to J do

12 Partition the Euler path E_i into $ST_i = \left\lceil \frac{(L(E_i))}{Vt} \right\rceil$ sub-paths of equal length and deploy two data mules at two ends of all the sub-paths.

13 end

14 Two data mules from each sub-path will move inward direction in synchronous, when they meet each other then reverse their movement directions to outward until they reach their starting positions again and continue the same process repeatedly.

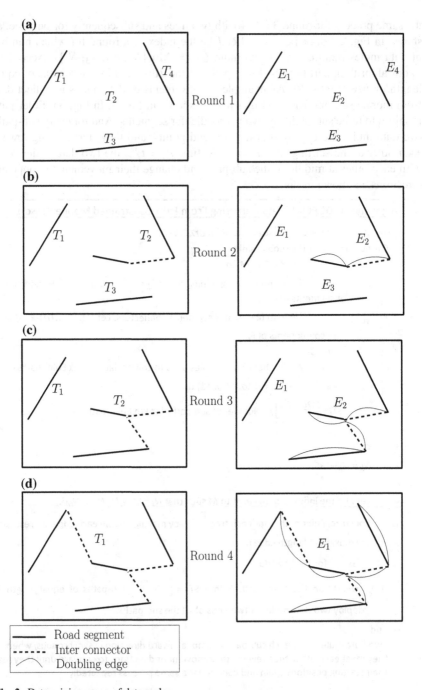

Fig. 2 Determining usage of data mules

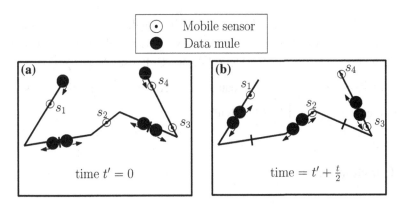

Fig. 3 Determining the paths of the data mules

4.2 Analysis

Theorem 2 *Algorithm DGPCMS ensures that each mobile sensor is visited by a data mule at least once in every t time period.*

Proof Let at time t, a mobile sensor node can be in one of the sub-paths which is periodically traversed by its two corresponding data mules. Since the length of the sub-path is at most Vt, it is bounded by two data mules from two ends. The mobile sensor cannot escape the sub-path without visited by its corresponding two data mules. All points of the sub-paths are also visited by data mules within t time interval. Hence, the theorem is proved.

Theorem 3 *Time complexity of the algorithm DGPCMS is $O(M^2 \log M)$.*

Proof DGPCMS algorithm determines a minimum length forest of k trees where k varies from 1 to M. It also determines the upper bound of the number of mules required to traverse Euler graph corresponding to the k trees. Therefore, time complexity from Step 1 to Step 8 is $O(M^2 \log M + M^2)$. Step 9 finds the minimum among M values, which can be determined in $O(M)$ time. Step 10 constructs Euler paths from all trees by doubling each edge of the trees that can be done on $O(M)$ time. The time complexity for Step 11 to Step 13 takes $O(M)$ time. Hence, the total time complexity of DGPCMS is $O(M^2 \log M)$.

Theorem 4 *Number of data mules used in algorithm DGPCMS is $\leq 4OPT$ (OPT denotes the minimum number of data mules require to collect data from all mobile sensors).*

Proof In the worst case, the data mules together must visit the full spanning forest within time period t. Since the speed of the data mule is V, therefore, the numbers of data mules required to collect the data from the mobile sensors

OPT $\geq \sum_{i=1}^{J} \left\lceil \frac{L(T_i)}{Vt} \right\rceil$, where $L(T_i)$ denotes the length of the tree T_i and according to the algorithm DGPCMS, J is the index of the round for which the number
of data mules used is minimum. Algorithm DGPCMS uses $N = 2 \sum_{i=1}^{J} \left\lceil \frac{L(E_i)}{Vt} \right\rceil \leq$
$2 \sum_{i=1}^{J} \left\lceil \frac{2L(T_i)}{Vt} \right\rceil \leq 2 \sum_{i=1}^{J} \left\lceil \frac{L(T_i)}{Vt} \right\rceil$ data mules to collect data from all mobile sensors.
Therefore, $N \leq 4 \sum_{i=1}^{J} \left\lceil \frac{L(T_i)}{Vt} \right\rceil \leq 4$OPT.

5 Conclusions

In this paper, we have proposed an approximation algorithm for data gathering protocol from mobile sensors using mobile data sinks. Our proposed algorithm overcomes
the limitation in [1]. The algorithm will return a solution within 4-factor of the optimal solution which will run in $O(M^2 \log M)$ time. In future, we will extend the work
by relaxing the paths of the mobile sensors from line segments to an arbitrary region,
where the mobile sensors can move arbitrarily within the bounded region.

Acknowledgements This work is conducted under the SERB Project ECR/2016/001035, Department of science and technology, Government of India.

References

1. Gorain, B., Mandal, P.S.: Line sweep coverage in wireless sensor networks. In: Communication
   Systems and Networks (COMSNETS), India, pp. 1–6 (2014)
2. Singh, S.K., Kumar, P., Singh, J.P.: An energy efficient odd-even round number based data collection using mules in WSNs. In: IEEE International Conference on Wireless Communications
   Signal Processing and Networking (WiSPNET) (2016)
3. He, J., Ji, S., Pan, Y., Li, Y.: Constructing load-balanced data aggregation trees in probabilistic
   wireless sensor networks. Trans. Parallel Distrib. Syst. 25(7), 1681–1690 (2014)
4. Liu, J.S., Wu, S.Y., Chiu, K.M.: Path planning of a data mule in wireless sensor network using
   an improved implementation of clustering-based genetic algorithm. In: IEEE Symposium on
   Computational Intelligence in Control and Automation (CICA), pp. 30–37 (2013)
5. Wu, S.Y., Liu, J.S.: Evolutionary path planning of a data mule in wireless sensor network
   by using shortcuts. In: IEEE Congress on Evolutionary Computation (CEC), pp. 2708–2715
   (2014)
6. Kim, D., Uma, R.N., Abay, B.H., Wu, W., Wang, W., Tokuta, A.O.: Minimum latency multiple
   data MULE trajectory planning in wireless sensor networks. IEEE Trans. Mob. Comput. 13(4),
   838–851 (2014)
7. Gao, S., Zhang H.: Energy efficient path-constrained sink navigation in delay guaranteed wireless sensor networks. J. Netw. 5(6), 658–665 (2010)
8. Mai, L., Shangguan, L., Lang, C., Du, J., Liu, H., Li, Z., Li, M.: Load balanced rendezvous data
   collection in wireless sensor networks. In: IEEE International Conference on Mobile Ad-Hoc
   and Sensor Systems (2011)
9. Gao, S., Zhang, H., Das, S.K.: Efficient data collection in wireless sensor networks with path-
   constrained mobile sinks. IEEE Trans. Mob. Comput. 10(5), 592–608 (2011)

10. Yao, Y., Cao, Q., Vasilakos, A.V.: EDAL: an energy-efficient, delay-aware, and lifetime balancing data collection protocol for heterogeneous wireless sensor networks. IEEE/ACM Trans. Netw. **23**(3), 810–823 (2015)
11. Little, J.D.C., Murty, K.G., Sweeney, D.W., Karel, C.: An algorithm for the traveling salesman problem. Oper. Res. **11**(6), 972–989 (1963)

Localization in 3D Wireless Sensor Network

Sneha Mishra, Sunil Kumar Singh and Prabhat Kumar

Abstract Most of the wireless sensor networks (WSN) are randomly deployed in their objective terrains and therefore require a localization mechanism for the deployed nodes. Many researchers have considered WSN as a 2D deployment scenario in their research works. But when we come to reality, the actual scene is in 3D. 2D localization algorithms will not remain applicable in 3D scenario if the height/depth parameter of nodes' coordinate has significant variations. The outcome of 2D-based localization algorithms may mislead the further computations involved in WSN. The paper aims to calculate the virtual coordinates of the sensor nodes deployed in a 3D area consisting of a single large obstacle. In this proposal, we are considering the deployment space with a single large obstacle and a distributed approach with four beacon nodes for providing localization of the deployed nodes. The paper uses region division concept for locating their virtual coordinates with respect to the obstacle. Each node is responsible for computing its coordinates. The possibility of some nodes not receiving minimum number of required signals from the beacon nodes is also considered and hence one-time transmission of the calculated data for each sensor node is included in the algorithm.

Keywords Wireless sensor network · Three dimension · Beacon nodes
Region code · Virtual coordinates

S. Mishra (✉) · S. K. Singh · P. Kumar
Department of Computer Science and Engineering, National Institute
of Technology Patna, Patna, Bihar, India
e-mail: mishrasneha1993@gmail.com

S. K. Singh
e-mail: sunil.cse14@nitp.ac.in

P. Kumar
e-mail: prabhat@nitp.ac.in

© Springer Nature Singapore Pte Ltd. 2018
R. Chaki et al. (eds.), *Advanced Computing and Systems
for Security*, Advances in Intelligent Systems and Computing 667,
https://doi.org/10.1007/978-981-10-8183-5_8

121

1 Introduction

A WSN is the collection of a huge number of small sensor nodes deployed in a sensing field to sense target events. Sensor nodes can sense their local environment, compute them and send those sensed events periodically to their neighbouring nodes. The aggregated sensed data reaches to the BS via intermediate nodes using wireless transmission techniques [1]. The BS is the end point for nodes data and it works as a gateway for connecting to the outside world. The major applications of WSN are object tracking, military surveillances, health monitoring, disaster relief and habitat monitoring.

Localization is a process of locating a sensor node in a deployment space. Traditionally or we say commonly, Global Positioning System (GPS) is used for locating objects. GPS requires at least four satellites at any instance for providing coordinates information with standard references. Use of GPS in WSN is not a feasible option because it adds an extra hardware, consumes more energy, increases cost, etc.

Localization process in WSN attempts to calculate nodes' location information with less energy consumption and computational complexity. Distance between reference node and node to be localized is an important factor in localization process. Based on the approach to calculate distance, localization algorithms can be classified into two categories: (1) Range-based localization and (2) Range-free localization. Range-based localization algorithms require extra hardware and they use physical properties of radio signals to compute distance information. They are further sub-classified, in accordance to the parameters of physical properties, as Received Signal Strength Indicator (RSSI) [2], Angle of Arrival (AoA) [3], Time of Arrival (TOA) [4] and Time Difference of Arrival (TDoA) [3]. Range-free localization algorithms exploit the existing topology like connectivity information for calculating distance. DV-HOP [5] is a range-free localization algorithm which uses tradition distance vector packet for computation.

In this paper, we aim to compute virtual coordinates of sensor nodes while keeping in mind the various constraints of WSN. Virtual positions of nodes can be efficiently used in WSN [6]. The paper presents a distributed range-based localization scheme which computes the coordinates in two steps, i.e. initially x and y coordinates are calculated and then z coordinate is determined. The deployment space considered contains a single large object. The location of the obstacle with respect to any node selected has also been considered in the work. Each sensor node maintains a 4-byte counter in accordance with signals they receive from beacon nodes. This helps in obstacle localization for a node.

2 Related Work

Several localization algorithms have been proposed for locating nodes of WSN effectively and efficiently. Most of them essentially incorporate the concept of utilizing a

Multi-dimensional Scaling (MDS) for localization in a 2D scenario using only a few anchor nodes. Authors in [7] present set of localization algorithm with classification based on placement type, processing and mobility of sensor nodes. Anchor node-based conventional centroid localization algorithm was proposed in [8] for detecting the locations of unknown nodes in a 2D scenario. Later, the traditional centroid-based localization was further improved in [9]. A centroid theorem applied to the volume coordinate system for coordinate tetrahedron has been proposed to improve the accuracy of localization so that the algorithm is applicable for WSN set in a 3D space. Both of the algorithms proposed in [8, 9] require anchor nodes in a large quantity. However, their localization accuracies show an inverse relationship to the number of anchor nodes used. Han et al. [10] present a comprehensive survey with different classification criteria. They classify algorithms based on the mobility of nodes and anchors. The authors discuss various algorithms in 2D as well as in 3D scenarios.

The application of MDS for locating sensor nodes in WSN was proposed in [11]. It aimed to address the issues faced in an anisotropic topology and uneven terrain. The proposed solution consisted of an MDS-based distributed localization algorithm in which each node creates a local estimate of the network and later all such estimates are compared with true physical locations of anchors to correct the estimate. This process of iteration is applied to all other nodes to get a correct estimation of the network. The concept of local optimization is applied in order to reduce the error obtained in the process of localization. The major drawback of the proposed algorithm was that since it iterated a distributed algorithm on all of the sensor nodes, it resulted in a higher rate of energy consumption during the process of localization. A 2D localization algorithm based Hierarchical MDS (HMDS) was proposed in [12] which comprised of three stages, viz. cluster creation, within cluster localization and merging of local coordinates. The algorithm has a major drawback that rendered it unfeasible to map a global coordinate system from a local coordinate system in case any disjoint clusters were evident. A higher rate of error is observed in localization techniques that use the shortest path for estimating distances. An MDS-MAP algorithm was proposed in [13] that employed classical MDS for obtaining map nodes in WSN. The algorithm was further improved to formulate MDS-MAP (P) [14] and MDS-MAP (R). The authors in [15] use the concept of a cooperative algorithm with MDS-MAP (P) that considers patches of relative estimate obtained from each node, which are assembled to locate the sensor nodes. The algorithms are distributed in nature. They consider each node for the process of localization. Classical MDS is used in [16]. The algorithm presents a centralized approach to solve localization problem in 3D WSN. The algorithm requires a minimum of five anchor nodes for effective localization. Its accuracy is directly proportional to the number of anchors used. The authors in [17] propose a method for underwater WSN in which they transform the 3D scenario into a 2D one using non-degenerative projection. They have a strong prior assumption that nodes are aware of their depth. The concept of degree of coplanarity is used by authors in [18]. They amalgamate DV-HOP concept with coplanarity to increase the accuracy of localization.

3   Network Model and Assumptions

In our proposed methodology, static sensor nodes are deployed in 3D space. The network has two types of nodes based on resources they acquire, namely, static nodes or nodes and beacon nodes. Beacon nodes are self-location aware nodes. The base station is situated outside the deployment area. Figure 1 shows deployment scenario.

3.1   Assumptions

The methodology considers the following assumption regarding deployment area and other components:

- Base station and beacon nodes are resourceful.
- Base station is located outside the objective space.
- Network consists of heterogenous nodes.
- There is a single large obstacle in deployment area which causes loss of signal.

3.2   Energy Consumption Model

Sensor nodes perform various functions in their lifetime such as sensing, computation, receiving, transmission, etc. Sensing and computation require negligible energy compared to receiving a transmitting. In the proposed work, we are considering two commonly used energy models in WSN, namely, free space propagation and mul-

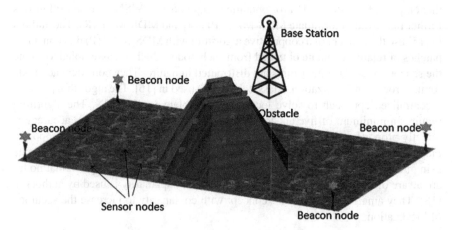

Fig. 1 Deployment scenario consider in algorithm

tipath fading channel model. Equation 1 represents mathematical representation of model. Transmission energy required to transmit p bits of data over a distance x is given by

$$E_T(p, x) = \begin{cases} \left(E_{el} * p + p * \varepsilon_{fs} * x^2\right) & x < d_0 \\ \left(E_{el} * p + p * \varepsilon_{mp} * x^4\right) & x \geq d_0 \end{cases} \tag{1}$$

where E_{el} is the electronic energy of sensor node for transmission of one bit of data. ε_{fs} is the amplifier energy for free space model. ε_{mp} is the amplifier energy for multipath fading model.

Threshold distance d_0 is given by

$$d_0 = \sqrt{\frac{\varepsilon_{fs}}{\varepsilon_{mp}}} \tag{2}$$

Energy consumed in receiving p bits of data is given by

$$E_R(p) = (E_{el} * p) \tag{3}$$

4 Proposed Work

Localization problem in WSN deals with providing location information to most of the nodes with less computational and communicational complexity. In this methodology, we propose a distributed range-based localization scheme for generating nodes location information. Some mathematical prerequisites are required which are discussed in Sect. 4.1. Section 4.2 explains the idea of region division. Proposed strategy to solve localization is mentioned in Sect. 4.3.

4.1 Mathematical Prerequisites

– RSSI ranging principle [19]:
  Given P_t is the transmitted power strength of signal, d is the distance between transceiver and receiver, and n is the transmission factor which depends on environmental factors, received power strength of signal (P_r) can be computed as

$$p_r = p_t \left(\frac{1}{d}\right)^n \tag{4}$$

– Given two points $A(x, y)$ and $B(a, b)$, distance (D_{AB}) between A and B will be

$$D_{AB} = p\,(x - a)^2 + (y - b)^2 \tag{5}$$

– Given three sides of a triangle a, b and c, an angle of triangle will be

$$A = \arccos\left(\frac{b^2 + c^2 - a^2}{2bc}\right) \tag{6}$$

4.2 Region Identification

The deployment space consists of several regions and each region represents availability of anchors signals. Regions are identified by a 4-byte counter (B_1, B_2, B_3, B_4) where value at each byte represents the strength of signal received by the node from respective beacon. For example, suppose region code of node A is (0, 2, 3, 4), it means strength of signal received from B_2 by node is 2 dBm. In Fig. 2, polygon bounded with red boundary represents the region where nodes receive signal only from beacon node 1 and 4. Figure 2 ignores reflection and diffraction of signals for simplicity of visualization.

Each node broadcasts its counter value. Node maintains the counter values of its neighbour node which can help to determine its neighbour location with obstacle.

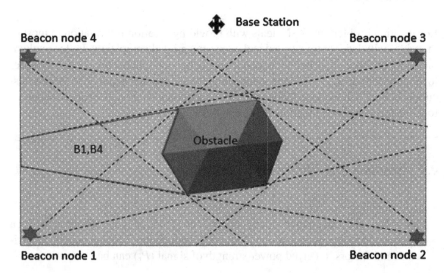

Fig. 2 Top view of deployment space

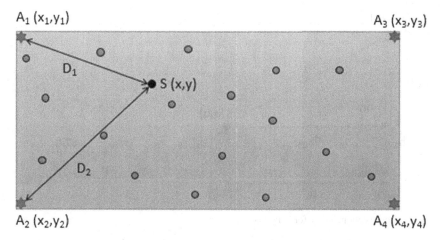

Fig. 3 Computation of *x* and *y* coordinates

4.3 Abscissa and Ordinates Computation

Nodes which receive location information of at least two nodes can compute their *x* and *y* coordinates using Eq. (5) as shown in Fig. 3. After solving quadratic equations, nodes get two different values among which one will lie inside deployment space, which will be the coordinate value for that node.

4.4 Height Computation

Height determination of nodes requires signal from single localized node. A trigonometric identity is used as mentioned in Eq. (6). Figure 4 shows a scenario in which sensor node is at height h_2 and beacon node is placed at height h_1, and height can be determined as

$$h_2 = h_1 - (BA) \tag{7}$$

4.5 Algorithm for Localization

The proposed algorithm is executed individually by each deployed sensor node. Four beacon nodes are used to cover the entire space. Beacon nodes are placed at same height. The algorithm divides the deployment area into regions, and each region represents a set of anchors whose signals are available in that area. Nodes which are in shadowing area (because of obstacle) of an anchor did not receive signal from

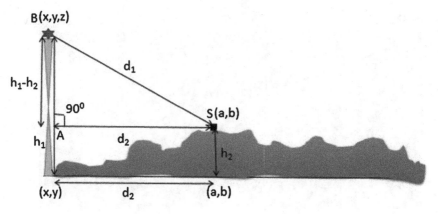

Fig. 4 Cross-sectional view for height computation

respective anchor. Nodes maintain the record of received signals which will use to determine their region. Nodes with minimum two anchors signal can compute their x and y coordinates. After that z coordinate is calculated. The proposed scheme is present in Algorithm 1.

Result : Coordinates of nodes
initialization;
Base station send start signal to beacon nodes;
Beacon nodes broadcast their location information;
while *All nodes get localized* **do**
  instructions;
  if *Node receive signal from atleast 2 beacons* **then**
      Node maintains a 4-byte counter, C, with respect to received
        signal;
      Node computes abscissa and ordinates, as mentioned in section 4.3;
      Node calculate z-coordinate, as mentioned in section 4.4;
      Broadcast the coordinates and value of C;

  else
      Node wait for signal;
  end
end

Algorithm 1: Proposed Localization Algorithm

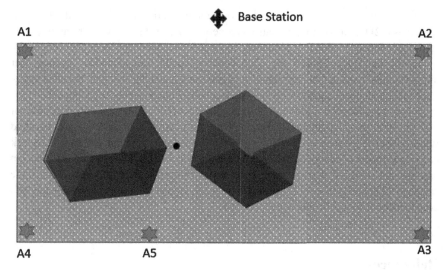

Fig. 5 Real-world scenario

5 Analysis

The proposed work considers a single large obstacle. However, it can also be applied to real-life scenarios which may not necessarily abide by the restriction of posing a single large obstacle. Figure 5 depicts a scenario where a sensor can be located in such a way that it is covered by multiple large obstacles. This hinders the exposure of the sensor node to the transmission ranges of the beacon nodes placed at location A1, A2, A3 and A4. In such cases, the beacon nodes can be provided with peripheral mobility, i.e. they can move from one point to another in their periphery. This shall ensure that the sensor node lies within the coverage area of the beacon nodes. In Fig. 5, the anchor node, initially placed at A4, is not able to provide appropriate coverage to the sensor node due to its location behind the large obstacle. However, after moving it from its initial location at A4 to a new location at A5 on the same periphery, the anchor node is able to cover the sensor node in its transmission range. The proposed approach has the advantage of performing localization even at a minimal supplement of two beacon nodes. Thus, the proposed approach requires only four mobile beacon nodes for conducting effective localization of the sensor nodes even in the areas having multiple large obstacles.

6 Conclusions and Future Works

The paper presents a distributed algorithm which considers 3D deployment space consisting of a single large obstacle. Most of the work done in WSN localization ignores height/depth attribute of nodes coordinate, i.e. they consider the deployment region as a 2D area which may cause erroneous result if taken in other calculations. Each node is responsible for calculating its coordinates which reduces the complexity of algorithm. The proposed approach for the localization of nodes requires four anchor nodes to maximize reachability of beacon signals. Nodes send their data only once throughout the algorithm flow which results in less energy consumption.

Number of beacons nodes can be reduced with some more efficient placement of beacons or at the cost of extra hardware. Communication load will be better handled with some efficient way to determine if the sent data has significance or not. The work can be extended to design a routing scheme [20] for WSN. Region code information can also be used for determining some useful inference regarding network. Since security- [21] and privacy [22]-related issues are also crucial in localization, we will consider them in our future works.

References

1. Akyildiz, I.F., Su, W., Sankarasubramaniam, Y., Cayirci, E.: Wireless sensor networks: a survey. Comput. Netw. **38**(4), 393–422 (2002)
2. Zheng, J., Wu, C., Chu, H., Xu, Y.: An improved rssi measurement in wireless sensor networks. Procedia Eng. **15**, 876–880 (2011)
3. Boukerche, A., Oliveira, H.A., Nakamura, E.F., Loureiro, A.A.: Localizationsystems for wireless sensor networks. IEEE Wirel. Commun. **14**(6), 6–12 (2007)
4. Ward, A., Jones, A., Hopper, A.: A new location technique for the active office. IEEE Pers. Commun. **4**(5), 42–47 (1997)
5. Niculescu, D., Nath, B.: Dv based positioning in ad hoc networks. Telecommun. Syst. **22**(1–4), 267–280 (2003)
6. Anwit, R., Kumar, P., Singh, M.: Virtual coordinates routing using vcp-min wireless sensor network. In: 2014 International Conference on Computational Intelligence and Communication Networks (CICN). IEEE, pp. 402–407 (2014)
7. Kumar, P., Singh, M., Triar, U., Anwit, R.: Localization system for wireless sensor network
8. Bulusu, N., Heidemann, J., Estrin, D.: Gps-less low-cost outdoor localization for very small devices. IEEE Pers. Commun. **7**(5), 28–34 (2000)
9. Chen, H., Huang, P., Martins, M., So, H.C., Sezaki, K.: Novel centroid localization algorithm for three-dimensional wireless sensor networks. In: 2008 4th International Conference on Wireless Communications, Networking and Mobile Computing. IEEE, pp. 1–4 (2008)
10. Han, G., Xu, H., Duong, T.Q., Jiang, J., Hara, T.: Localization algorithms of wireless sensor networks: a survey. Telecommun. Syst. **52**(4), 2419–2436 (2013)
11. Ji, X., Zha, H.: Sensor positioning in wireless ad-hoc sensor networks using multidimensional scaling. In: INFOCOM 2004. Twenty-third Annual Joint Conference of the IEEE Computer and Communications Societies, vol. 4. IEEE, pp. 2652–2661 (2004)
12. Yu, G.-J., Wang, S.-C.: A hierarchical mds-based localization algorithm for wireless sensor networks. In: 22nd International Conference on Advanced Information Networking and Applications (AINA 2008). IEEE, pp. 748–754 (2008)
13. Shang, Y., Ruml, W., Zhang, Y., Fromherz, M.P.: Localization from mere connectivity. In: Proceedings of the 4th ACM International Symposium on Mobile Ad Hoc Networking & Computing. ACM, pp. 201–212 (2003)
14. Shang, Y., Ruml, W.: Improved mds-based localization. In: INFOCOM 2004. Twenty-third Annual Joint Conference of the IEEE Computer and Communications Societies, vol. 4. IEEE, pp. 2640–2651 (2004)

15. Shang, Y., Meng, J., Shi, H.: A new algorithm for relative localization in wireless sensor networks. In: Proceedings of 18th International Conference on Parallel and Distributed Processing Symposium, 2004. IEEE, p. 24 (2004)
16. Chaurasiya, V.K., Jain, N., Nandi, G.C.: A novel distance estimation approach for 3d localization in wireless sensor network using multi dimensional scaling. Inf. Fusion **15**, 5–18 (2014)
17. Teymorian, A.Y., Cheng, W., Ma, L., Cheng, X., Lu, X., Lu, Z.: 3d under water sensor network localization. IEEE Trans. Mob. Comput. **8**(12), 1610–1621 (2009)
18. Xu, Y., Zhuang, Y., Gu, J.-J.: An improved 3d localization algorithm for the wireless sensor network. Int. J. Distrib. Sens. Netw. **2015**, 98 (2015)
19. Xu, J., Liu, W., Lang, F., Zhang, Y., Wang, C.: Distance measurement model based on rssi in wsn. Wirel. Sens. Netw. **2**(08), 606 (2010)
20. Kumari, P., Singh, M.P., Kumar, P.: Survey of clustering algorithms using fuzzy logic in wireless sensor network. In: 2013 International Conference on Energy Efficient Technologies for Sustainability, pp. 924–928 (April 2013)
21. Abhishek, K., Roshan, S., Kumar, P., Ranjan, R.: A Comprehensive Study on Multifactor Authentication Schemes. Springer, Berlin, Heidelberg, pp. 561–568 (2013). http://dx.doi.org/10.1007/978-3-642-31552-757
22. Roy, P.K., Singh, J.P., Kumar, P., Singh, M.: Source location privacy using fake source and phantom routing (fsapr) technique in wireless sensor networks. Procedia Comput. Sci. **57**, 936–941 (2015). http://www.sciencedirect.com/science/article/pii/S1877050915020153

Part IV
Security Systems

EAER-AODV: Enhanced Trust Model Based on Average Encounter Rate for Secure Routing in MANET

Saswati Mukherjee, Matangini Chattopadhyay, Samiran Chattopadhyay and Pragma Kar

Abstract Secure routing is an integral part of MANET for protecting the network operations from malicious behaviour. However, the dynamic nature of MANET demands some stringent mechanism that can ensure security in routing. We have redefined an existing trust model by forming a 3-dimensional metric in order to deal with the dynamic topology of MANET. In this paper, we have proposed a trust based routing protocol named Enhanced Average Encounter Rate-AODV (EAER-AODV) that employs the trust model based on nodes' opinion. In EAER-AODV, opinion represents the trust among nodes which is updated frequently according to the protocol specification. Trust based on recommendation is also used to exchange the trust information among nodes. In this protocol, a node selects a routing path according to the trust values of its neighbour nodes. Extensive simulation analysis shows that EAER-AODV can effectively avoid the malicious nodes and nodes having frequent mobility while selecting the routes. It is also shown that EAER-AODV is compared with existing methods to prove its efficacy.

Keywords Distrust · Trust · Mobility · Average encounter · Routing

1 Introduction

MANET is a dynamic topology where nodes form a network without any fixed network infrastructure. However, lack of centralized control and dynamic topology impose a challenge on the security mechanism in MANET. Malicious nodes often participate in misrouting the packets and high mobility nodes disrupt the packet

S. Mukherjee (✉) · M. Chattopadhyay · P. Kar
School of Education Technology, Jadavpur University, Kolkata, India
e-mail: saswatimuk@gmail.com

M. Chattopadhyay
e-mail: matanginic@gmail.com

S. Chattopadhyay
Department of Information Technology, Jadavpur University, Kolkata, India

© Springer Nature Singapore Pte Ltd. 2018
R. Chaki et al. (eds.), *Advanced Computing and Systems for Security*, Advances in Intelligent Systems and Computing 667, https://doi.org/10.1007/978-981-10-8183-5_9

transmission. Thus, it is necessary to secure the existing routing protocols in MANET. There are several works that have used cryptography-based mechanisms to defend against external attacks; however, protecting the network from internal attacks is still to be addressed properly. To defend against internal attacks, trust based mechanism plays a significant role in network operations. The concept of trust is derived from subjective probability which states that, an agent has some expectation about another agent's future action [1].

In this paper, we have built a trust mechanism by considering two major factors for direct trust evaluation: Successful Cooperation Frequency (SCF) and Average Encounter Rate (AER). To create a stable network topology, we have utilized the concept of average encounter rate while calculating nodes' direct trust. Moreover, modified Dempster-Shafer (D-S) theory is used to combine multiple pieces of trust evidence and compute the recommended trust. By extending the AODV protocol, we put forward a trust based secure routing protocol in MANET, referred to as *EAER–AODV*. The trust model considers the opinion of a node which is represented as a 3-dimensional metric. The trust model selects the most reliable route from a set of several routing paths for packet transmission. During route selection, nodes which drop packets and frequently change their locations are discarded. Extensive ns-2 simulations are performed that demonstrate the efficacy of our work compared to the existing ones. Analysis shows that our method has outperformed the existing ones in terms of network throughput, packet delivery ratio, average end-to-end delay and routing overhead. This work is a significant extension of AER-AODV [2]. The salient improvements are discussed at the end of Sect. 2.

The rest of this paper is organized as follows. Section 2 presents the related works. In Sect. 3, the description of trust model is given. Section 4 presents the proposed trust based secure routing protocol. Performance analysis is presented in Sect. 5. Finally we conclude the paper in Sect. 6.

2   Related Works

The concept of using trust and selecting a trusted route for packet transmission is an important aspect in Ad hoc networks. The objective of trust establishment mechanism is to identify malicious nodes in the path of routing and to prevent the adversaries from advertising themselves as good. Several researchers have proposed various trust evaluation models for secure routing.

In [3], a source node establishes a forwarding route and a backup route using hop counts and trust values. The path which has the maximum trust value is selected for routing. D-S theory is applied for computing the recommended trust. This algorithm improves packet delivery rate and end to end delay. However, the routing overhead is not analyzed in the presence of malicious nodes.

Cho et al. [4] proposed a distributed trust-based public key management approach for MANETs using a soft security mechanism and trust threshold. In this approach, each node make local peer-to-peer trust assessment for distributed decision making based on a composite trust metric. The optimal trust threshold is applied at runtime for differentiating trustworthy versus untrustworthy nodes. The method considers a single threshold for node trustworthiness classification and assessment of indirect trust is not considered.

Liu et al. [5] proposed a trust model where each node communicates with their neighbours for obtaining the direct trust or combining the recommended trust. Intermediate hops are selected on the basis of high trust value. However, the algorithm fails to defend against bad mouthing attack and colluding malicious nodes.

In [6], a dynamic trust mechanism using trust updates and collaborative filtering is introduced. The trust updating algorithm observes the fluctuation of trust values and also considers the time aging, penalty and reward factors. This mechanism eliminates fake recommended trust by applying collaborative filtering. However, this mechanism cannot detect dishonest recommendations.

Malicious nodes are detected using a guard node in AODV protocol [7]. The trust of each node is calculated for route selection and a predefined threshold is used to detect malicious nodes. However, this algorithm suffers from bad mouthing attack and energy constraints.

The behaviour of a node is observed for specific time duration in [8] and the trust is computed using the behaviour summary. A Single Intrusion Detection (SID) packet is used to broadcast and inform all participating neighbours about the malevolent activity in the network. The method requires huge memory for storing the reputation of nodes.

In [9], a node-based trust management (NTM) scheme is introduced which considers a mobile agent system that runs on each node. NTM consists of three components: Node Initiators (NIs), Trust Monitors (TMs) and Trust Evaluators(TEs). The scheme uses symmetric keys to encrypt the aggregated trust and reputation. However, the scalability remains as a scope for further experimentation.

Pissinou et al. [10] designed a Trust embedded AODV (T-AODV) protocol that establishes a secure end-to-end path using nodes' collaboration. An extra trust-level field is incorporated in the header of RREQ packet. This trust field is modified when RREQ packet is received by an intermediate node. Route is selected using the trust level of a third party. T-AODV protocol cannot deal with colluding malicious nodes.

TAODV is an extension of AODV routing protocol that protects routing behaviors in MANETs [11]. In TAODV, the trust among nodes are based on opinion. Each node can flexibly define its own opinion threshold during cryptographic operations. The opinions are dynamic and updated frequently. Computational overhead is less in this scheme; however, this scheme is slow to detect malicious nodes.

Sarkar and Datta [12] proposed a trust module that computes and compares the trust using nodes' mobility. Their trust evaluation consists of three phases: initialization phase, update phase and re-establish phase. However, this scheme does not consider the malicious behaviour of packet dropping during routing.

To provide a secure routing scheme, trust-based reputation system and multipath routing are proposed in [13] in conjunction with soft encryption system. However, an algorithm is required to select a reliable route from a set of routes. Moreover, the routing scheme needs to be adapted to be used in AODV protocol.

The limitations of the existing solutions as discussed motivated us to propose a trust based secure routing in AODV protocol. We have used the concept of AER [14] to handle the mobility factor which eventually mitigates the issue of slow response in detecting the malicious nodes. Moreover, the use of recommended trust helps to defend against bad mouthing attack and dishonest recommendation. Furthermore, we have addressed the issue of routing overhead, packet dropping behaviour of nodes, and scalability in our method. The proposed method is being implemented in AODV protocol for analyzing and improving the packet delivery ratio, throughput and end-to-end delay. Our scheme does not rely on storing nodes' reputation, thus, memory requirement is nominal.

The proposed work EAER-AODV is an extension of AER-AODV [2]. In [2], the trust model considers the direct trust and the recommended trust. D-S theory is used to compute recommended trust without specifying the range of uncertainty. In this paper, we have redefined the trust model based on opinion which is represented as a 3-dimensional metric. The 3-dimensional metric consists of trust, distrust and uncertain components. The trust and uncertain components specify the trust threshold and the range of uncertainty explicitly. Consequently, the algorithm is modified according to the definition of the trust model. The simulation is conducted using NS-2 and hence results are more authentic and reliable. Packet delivery ratio and throughput with respect to malicious nodes are considered to verify the efficacy of the algorithm in [2]. However, in this work, the algorithm is examined and validated considering many additional performance metrics, i.e., packet delivery ratio, throughput and routing overhead with respect to malicious nodes and varying speed. Furthermore, EAER-AODV is compared with TDS-AODV [3] and TAODV [11] methods to prove its efficacy.

3 Trust Model

The concept of AER and SCF as reported in [2] is further extended and utilized in this paper for calculating the direct trust of a node.

Trust Definition: The definition of 'trust' refers to the belief that one node holds about another, based on past experiences, knowledge of entity behavior and/or recommendations from trusted entities [15]. In our work, 'trust' refers to the opinion of a node i regarding a node j about the ability to perform the network operations successfully. In this work, we have used 'trust' and 'opinion' synonymously. Each node has its opinions about some other nodes trustworthiness, which are obtained by directly communicating with other nodes or by combining other nodes recommendations. In our trust model, the opinion (ξ) of a node i regarding a node j is a 3-dimensional metric that consists of three components: trust (t), distrust (d) and uncertain (u) and is defined as follows.

$$\xi_{i,j} = t_{i,j} + d_{i,j} + u_{i,j} = 1 \qquad (1)$$

In Eq. (1), trust $t_{i,j}$ means the probability that the node i can trust the node j and distrust $d_{i,j}$ refers to the probability that the node i cannot trust the node j. Uncertainty $u_{i,j}$ refers to the interval between trust and distrust components and sum of these three components is 1.

Trust $t_{i,j}$ is defined as the direct trust $dt_{i,j}$ of the node i regarding the node j. The direct trust $dt_{i,j}$ is computed by evaluating two trust factors, i.e., AER and SCF described in the next subsection. Distrust $d_{i,j}$ becomes zero when the computed direct trust $dt_{i,j}$ is less than β (value of β is mentioned in Sect. 3.4).

Uncertainty $u_{i,j}$ is defined as the recommended piece of information obtained from another node m about the node j. The recommended piece of information is then combined and evaluated using modified Dempster-Shafer theory that generates the recommended trust $rt_{i,j}$.

3.1 Average Encounter Rate

In MANET, nodes which are more stable are considered for trust establishment. It is observed in the literature [3] that nodes having less mobility can successfully transmit packets and thus, are more trusted and reliable. A node, which changes its position frequently, is often involved in unsuccessful network operations. The encounter rate [14] denotes a set of new encounters (nodes) which a mobile node j experiences in duration $T(t_i, t_{i+1})$ moving at velocity v. The node j that experiences an average number of new nodes or encounters per unit time T is defined as the average encounter rate of a node j, i.e., AER_j.

$$AER_j = |E_j| \div T \qquad (2)$$

where $|E_j|$ is the no. of encounters or nodes of the set E_j. Conceptually, higher AER_j value denotes that node j frequently changes its location which at times results in link failure. Thus, node j is considered to be less trusted and reliable.

3.2 Successful Cooperation Frequency (SCF)

Successful network operations depend on multiple parameters such as no. of packets forwarded (pf), the number of packets dropped (pd) and the number of false packets injected (pw). These parameters are utilized for calculating SCF of each node. Each observing node i predicts the trust of each node j by using $\text{SCF}_{i,j}$ as given in the following equation.

$$\text{SCF}_{i,j} = \frac{\text{pf}_{i,j}}{\text{pf}_{i,j} + \text{pd}_{i,j} + \text{pw}_{i,j}} \tag{3}$$

Direct trust $dt_{i,j}$ is the sum of weighted AER_j and weighted $\text{SCF}_{i,j}$ as given in the following equation.

$$dt_{i,j} = \left(\omega_1 \times \text{AER}_j\right) + \left(\omega_2 \times \text{SCF}_{i,j}\right) \tag{4}$$

where $0 < \omega_1 \le 1$ and $0 < \omega_2 \le 1$ are weights assigned to each trust factor, AER and SCF. These weights are adjusted according to the successful and unsuccessful network operation of nodes.

3.3 Modified D-S Evidence Theory for Recommended Trust

In building a trust model, it is often difficult to obtain an accurate observation and estimation of mobile nodes. To deal with uncertain information, probability theory is not useful. Dempster Shafer theory [16] is a mathematical approach that handles uncertain knowledge. Evidence from different sources are collected to arrive at a degree of belief which is represented by a belief function. In D-S theory, a set of all possible mutually exclusive propositions are represented in the frame of identification Θ [15]. In our work, a set of all possible mutually exclusive propositions are defined as $\{\{\emptyset\}, \{T\}, \{-T\}, \{T, -T\}\}$ where $\emptyset$ is the empty set and $\{T\}, \{-T\}, \{T, -T\}$ represent the proposition of nodes as "Trust", "Distrust" and "Uncertainty" respectively. Weights are assigned to the evidences that support the proposition of nodes. Each node i contributes its opinion about a target node j by assigning belief over Θ. Basic Probability Function (BPA) is the assignment function or the Mass Function $m : 2^{\Theta} \rightarrow [0, 1]$ of node i denoted as m_i. The 'confidence interval' is the probability that a node i trusts a node j and it is represented by Belief $\text{Bel}_i(T)$ and Plausibility $\text{Pl}_i(T)$. BPA, which satisfies the following equations.

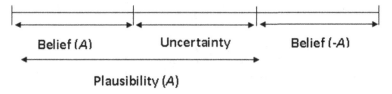

Fig. 1 Belief and plausibility

$$\Sigma m(A)|A \subseteq \Theta = 1, \quad m(\varnothing) = 0 \tag{5}$$

where A is a member in the frame of identification; $m(A) > 0$ is the basic confidence level of A, which represents the proportion of all evidence that supports the element A. The certainty associated with a given member A is defined by Belief Interval. Belief Interval is the difference between Belief and Plausibility. It also represents the range of maximum uncertainty. The association between Belief and Plausibility is shown in Fig. 1.

The belief function is the sum all evidences A that support the given proposition is 'Trusted $\{T\}$'. The belief function is defined as follows:

$$\text{Bel}_i(T) = \sum_{A \subseteq T} m(A), \quad \forall A \subseteq \Theta \tag{6}$$

where A denotes all evidence of node i observation that supports the target node j is 'Trusted'.

Plausibility function is the sum of all observations that intersect with the proposition 'Trusted':

$$\text{Pl}_i(T) = 1 - \text{Bel}_i(-T), \quad \forall A \subseteq \Theta \tag{7}$$

Thus, Plausibility $\text{Pl}_i(T)$ can also be represented by the following equation:

$$\text{Pl}_i(T) = 1 - d_{i,j} \tag{8}$$

Let the observation of node i is m_i and the observation of node j is m_j for each proposition, say, 'Trusted $\{T\}$'. Then, modified D-S rule is applied to combine the independent pieces of observation m_i and m_j to get the opinion about the target node as shown here.

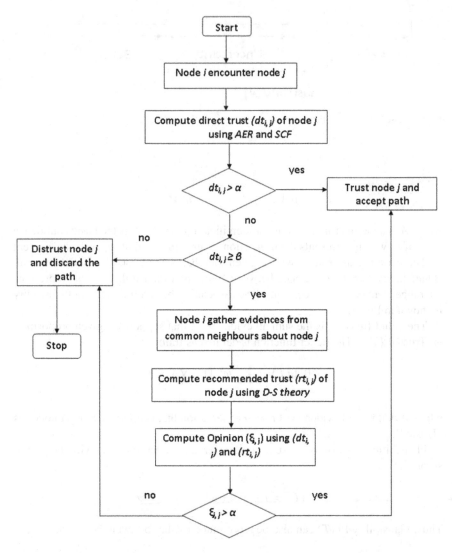

Fig. 2 Flowchart of trust evaluation mechanism

$$m_1 \oplus m_2(T) = \frac{\sum_{A \cap A_{k'} = A} m_i(A_k) m_j(A_{k'})}{\sum_{A \cap A_{k'} = \emptyset} m_i(A_k) \cdot m_j(A_{k'})} \tag{9}$$

Let ω_i and ω_j are the respective weights assigned to m_i and m_j. It is evident that nodes which perform network operations successfully are assigned the highest weight and nodes which frequently change their position are assigned the lowest weight. The weighted pieces of observation are combined to calculate the recommended trust of the target node as given here.

$$u_{i,j} = m_1 \oplus m_2(T) = \frac{\sum_{A \cap A_{k'} = A} \left[\omega_i m_i(A_k) \cdot \omega_j m_j(A_{k'}) \right]}{\sum_{A \cap A_{k'} = \emptyset} \left[\omega_i m_i(A_k) \cdot \omega_j m_j(A_{k'}) \right]} \tag{10}$$

According to our trust model, $u_{i,j} = rt_{i,j}$.

3.4 Opinion Evaluation

The overall trust or opinion is the sum of direct observation of a node i about a node j, i.e., $dt_{i,j}$ and recommended observation of the neighboring node m about the target node j, i.e., $rt_{i,j}$. The distrust component as mentioned in Eq. (1) is not directly considered in Eq. (11). The overall trust or opinion $\xi_{i,j}$ is calculated as follows.

$$\xi_{i,j} = \begin{cases} dt_{i,j}, & \text{if } dt_{i,j} > \alpha \\ dt_{i,j} + rt_{i,j}, & \text{if } \beta \leq dt_{i,j} < \alpha \\ 0, & \text{if } dt_{i,j} < \beta \end{cases} \tag{11}$$

We have chosen α to be 0.7 and β to be 0.5 in this paper.

4 Trust-Based Secure Routing Protocol

To have a clear understanding of the proposed approach, a flowchart of the proposed method is given in Fig. 2. The trust-based secure routing includes two processes: (i) Route Discovery and (ii) Trusted Route Selection.

4.1 Route Discovery

Let the source node and destination node be i and j respectively.

Step 1: In route discovery phase, the node i generate a *RREQ* packet which is broadcast to all its neighbours N_i.

Algorithm 1: Process of selecting trusted path

Input: Set of distinct routing paths (Ω) between the source and the destination
Output: Trusted path p_n between the source and the destination

1: **for** each path $p_n \in \Omega$ **do**
2:     Set *flag*=1
3:         **for** each node $i \in p_n$ **do**
4:             Node i computes $dt_{i,j}$ of next hop j using weighted AER weighted
    SCF
5:             **if** $dt_{i,j} > \alpha$ **then**
6:                 Node i regard node j as 'Trusted'
7:             **else**
8:                 **if** $dt_{i,j} \geq \beta$ and $dt_{i,j} < \alpha$ **then**
9:                     Node i treats node j as 'Uncertain'
10: Node i computes opinion $\xi_{i,j}$ of node j using direct trust $dt_{i,j}$ and recommended
    trust $rt_{i,j}$
11:                     **if** $\xi_{i,j} \geq \alpha$ **then**
12:                         Node i regard node j as 'Trusted'
13:                     **else**
14: Node i regard node j as 'Distrust' and the path p_n is rejected.
15:                         Set *flag*=0
16:                         break
17:                     **end if**
18:                 **else**
19:                     Node i regard node j as 'Distrust' and the path p_n is rejected.
20:                     Set *flag*=0
21:                     break
22:                 **end if**
23:             **end if**
24:         **end for**
25:     **if** *flag*=1 **then**
26:         Path p_n is 'Trusted'
27:     **end if**
28: **end for**

Step 2:   Each node $n \in N_i$ checks previously received *RREQ* packet from the source
         node i.
Step 3:   If the *RREQ* packet is found to be duplicate, then it is rejected.
Step 4:   Otherwise, *RREQ* packets are accepted by n and a reverse route entry is
         assigned to the route table.

Step 5: Each *RREQ* packet is further transmitted to the neighbours of node *n* until it reaches the node *j*.

Step 6: Once the node *j* received the *RREQ* packet, it generates *RREP* packet.

Step 7: *RREP* packet travels in the reverse path through which the *RREQ* packet has reached the destination.

Thus, a forward route is established from source *i* to destination *j*.

4.2 *Trusted Route Selection*

Multiple paths exist between the source and the destination after the route is discovered. The discovered paths may include some malicious nodes and high mobility nodes. Thus, trusted route selection algorithm is required to identify and select those paths which do not have any malicious nodes and high mobility nodes. The direct trust and the recommended trust of each intermediate node in a path are evaluated to select a reliable next hop for secure packet transmission. The route selection algorithm starts with considering one path at a time from the set of multiple distinct paths for a given source and destination. Algorithm 1 presents the trusted route selection algorithm.

5 Results and Analysis

To verify the efficacy and performance of our method, experiments are conducted using NS-2.35 [17].

5.1 *Simulation Environment*

We have considered 70 mobile nodes in our simulation which are randomly placed in a 1000×1000 square field. All nodes move independently following random waypoint model and the speed is varied from 5 to 30 m/s. The transmission range of each node is set to 250 m. In our simulation, we have varied the number of malicious nodes from 5 to 30. When a packet is received by a malicious node, it either drops or selectively forwards the packet to a node. Thus, a malicious node behaves like a blackhole node. Such malicious behaviour is added to the AODV protocol to analyze the performance of our method. Random nodes are selected to perform the malicious activity by dropping packets. All experimental data are obtained by averaging 50 runs. The simulation parameters are enlisted in Table 1.

5.2 Performance Metrics

We have considered four metrics for measuring the performance of our proposed EAER-AODV: (i) Throughput: number of packets received by the destination node per unit time. (ii) Packet Delivery Ratio (PDR): the ratio of the number of packets received to the total number of packets. (iii) Average end-to-end delay: time interval between packet transmission and packet reception and (iv) Routing Overhead: number of routing packets received divided by the total number of data packets.

The proposed method EAER-AODV is compared with the state-of-art of TDS-AODV [3] and TAODV [11] methods.

5.3 Simulation Results

Figures 3 and 4 depict the average throughput of EAER-AODV, TDS-AODV and TAODV with respect to different percentage of malicious nodes and varying speed. The average routing throughput of EAER-AODV is 49.75% higher than TDS-AODV and 33.75% higher than TAODV. Figure 3 indicates that EAERAODV can effectively prevent malicious nodes from becoming the next hop. In this experiment, the speed varies between 24 and 30 m/s.

As shown in Fig. 4, initially the throughput of EAER-AODV is lower than TDS-AODV and TAODV, but as the speed increases, it outperforms both TDS-AODV and TAODV. The reason is, nodes having higher AER value are prevented from becoming the next hop. Average throughput of EAER-AODV is 22% higher than TDS-AODV; however, the throughput of EAER-AODV is 14% lower than TAODV because of low throughput at low mobility scenario. This experiment is performed with 30 malicious nodes.

Table 1 Simulation parameters		
Channel type	Channel/Wireless channel
Radio-propagation model	Propagation/Two Ray Ground
MAC type	Mac/802 11
Link layer type	LL
Antenna model	Antenna/Omni Antenna
Simulation time	180 s
Traffic source	CBR
Number of nodes	70
Transmission radius	250 m
Mobility model	Random waypoint
Max speed	0–30 m/s
Number of malicious nodes	5, 10, 15, 20, 25, 30

Fig. 3 Network throughput
with different number of
malicious nodes

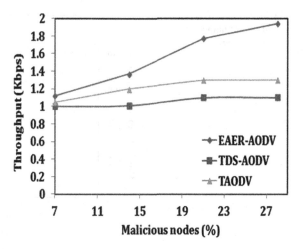

Fig. 4 Network throughput
at different speed

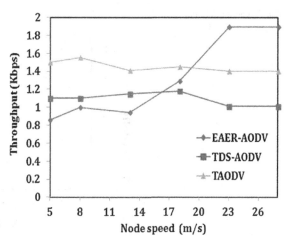

The packet delivery ratio (PDR) of EAER-AODV, TDS-AODV and TAODV is shown in Figs. 5 and 6. In Fig. 5, PDR of EAER-AODV rises slowly; however, as the number of malicious nodes increase, EAER-AODV shows an improvement over TDS-AODV and TAODV. The average PDR of EAER-AODV is 10.3% higher than TDS-AODV and 19% higher than TAODV. The reason is, only trusted intermediate nodes are selected for delivering packets to the destination.

In Fig. 6, with the increasing speed, the packet delivery ratio of EAERAODV shows an improvement over TDS-AODV and TAODV. The PDR of EAER-AODV is 8.5% higher than TDS-AODV and 21.7% higher than TAODV.

We present the average end-to-end delay comparisons of EAER-AODV, TDSAODV and TAODV in Figs. 7 and 8 considering 25 nodes with high mobility. In Fig. 7, the average delay of EAER-AODV drops faster than both TDSAODV

Fig. 5 Packet delivery ratio
with different number of
malicious nodes

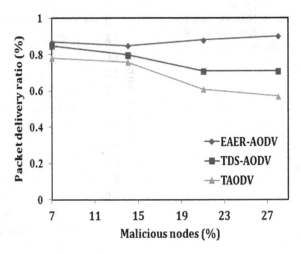

Fig. 6 Packet delivery ratio
at different speed

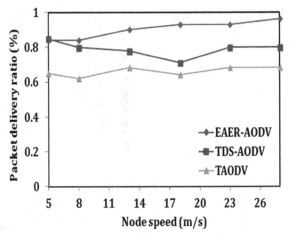

and TAODV because only the reliable and trusted hops are selected which reduces end-to-end packet transmission time. Figure 8 depicts the average delay of all three methods with varying speed. The delay rises with an increase in the maximum speed. However, the average delay of TDS-AODV is 47.83% higher than that of EAER-AODV and the average delay of TAODV is 54% higher than that of EAER-AODV. These results show that EAER-AODV has successfully captured the dynamic nature of the topology.

Figure 9 depicts the average routing overhead with respect to the number of malicious nodes. The average speed of nodes is set to 24 m/s. The routing overhead is the number of control packets transmitted per data packet delivered at the destination. If we keep the number of nodes fixed and vary the number of malicious nodes, routing overhead increases but the rate of increase is less.

Fig. 7 Average end-to-end
delay with different number
of malicious nodes

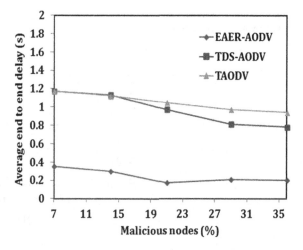

Fig. 8 Average end-to-end
delay at different speed

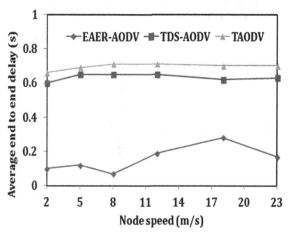

This is due to the fact that, in EAER-AODV, a significant number of control packets
is transmitted to establish a secure path from source to destination by considering
AER factor. Once the path is established by EAER-AODV, chances of link failure is
less.

In Fig. 10, the average routing overhead is shown with respect to varying speed
and different percentage of malicious nodes. In all four curves, the routing overhead
rises with increase in the speed because presence of high mobility nodes disrupts
the routing table frequently. Moreover, with the presence of different percentage of
malicious nodes, the rise is also significant. The reason is, if a malicious node comes
in the path of routing, it is eventually discarded. Therefore, selecting only the trusted
path from the several routing paths increases the routing overhead.

Fig. 9 Routing overheard
with varying malicious nodes

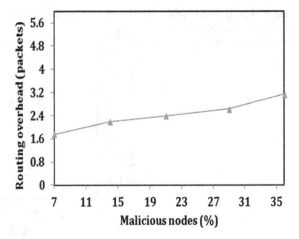

Fig. 10 Routing overheard
with varying node speed

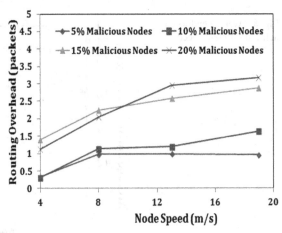

6 Conclusion

In our work, we have designed a trust mechanism after analyzing various trust
models for routing in ad hoc networks. Our trust model uses a node's opinion which
is a 3-dimensional metric that consists of trust, distrust and uncertainty components.
Trust component refers to the direct trust which is evaluated using average encounter
rate and successful cooperation frequency. Uncertainty component refers to the
recommended trust which is computed using modified D-S theory. Then, a trusted
route is discovered between the source and the destination using the trust model. The
source node selects the most reliable next hop for routing, thus, refusing high mobility
nodes and malicious nodes. Finally, we have validated the efficacy of EAER-AODV
by comparing its performance with TDS-AODV and TAODV. The simulation
results show that our method can effectively discard the malicious nodes and high

mobility nodes while building the route. Moreover, an improvement is observed in the throughput, packet delivery ratio, average end-to-end delay and routing overhead.

In future, this work can further be extended by studying and analyzing the energy consumption of each node and its trade-off with the security mechanisms.

References

1. Levi, M., Gambetta, D.: Trust: Making and Breaking Cooperative Relations (1991)
2. Mukherjee, S., Chattopadhyay, M., Chattopadhyay, S.: A novel encounter based trust evaluation for aodv routing in manet. In: Applications and Innovations in Mobile Computing (AIMoC). IEEE, pp. 141–145 (2015)
3. Feng, R., Che, S., Wang, X., Yu, N.: A credible routing based on a novel trust mechanism in ad hoc networks. Int. J. Distributed Sens. Netw. **2013** (2013)
4. Cho, J.H., Chen, R., Chan, K.S.: Trust threshold based public key management in mobile ad hoc networks. Ad Hoc Netw. **44**, 58–75 (2016)
5. Liu, Z., Lu, S., Yan, J.: Secure routing protocol based trust for ad hoc networks. In: Eighth ACIS International Conference on Software Engineering, Artificial Intelligence, Networking, and Parallel/Distributed Computing, 2007, vol. 1. IEEE, pp. 279–283 (2007)
6. Sancheng, P., Weijia, J., Guojun, W., Jie, W., Minyi, G.: Trusted routing based on dynamic trust mechanism in mobile ad-hoc networks. IEICE Trans. Inf. Syst. **93**(3), 510–517 (2010)
7. Raza, I., Hussain, S.A.: Identification of malicious nodes in an aodv pure ad hoc network through guard nodes. Comput. Commun. **31**(9), 1796–1802 (2008)
8. Chauhan, A., Patle, A., Mahajan, A.: A better approach towards securing mobile adhoc network. Int. J. Comput. Appl. **20**(8), 6–11 (2011)
9. Ferdous, R., Muthukkumarasamy, V., Sattar, A.: A node-based trust management scheme for mobile ad-hoc networks. In: 2010 4th International Conference on Network and System Security (NSS). IEEE, pp. 275–280 (2010)
10. Pissinou, N., Ghosh, T., Makki, K.: Collaborative trust-based secure routing in multihop ad hoc networks. In: International Conference on Research in Networking. Springer, pp. 1446–1451 (2004)
11. Li, X., Lyu, M.R., Liu, J.: A trust model based routing protocol for secure ad hoc networks. In: Proceedings of Aerospace Conference, 2004, 2004 IEEE, vol. 2. IEEE, pp. 1286–1295 (2004)
12. Sarkar, S., Datta, R.: A mobility factor based path selection scheme for mobile ad-hoc networks. In: 2012 National Conference on Communications (NCC). IEEE, pp. 1–5 (2012)
13. Narula, P., Dhurandher, S.K., Misra, S., Woungang, I.: Security in mobile adhoc networks using soft encryption and trust-based multi-path routing. Comput. Commun. **31**(4), 760–769 (2008)
14. Son, T.T., Le Minh, H., Sexton, G., Aslam, N.: A novel encounter-based metric for mobile ad-hoc networks routing. Ad Hoc Netw. **14**, 2–14 (2014)
15. Chatterjee, P., Sengupta, I., Ghosh, S.K.: A distributed trust model for securing mobile ad hoc networks. In: 2010 IEEE/IFIP 8th International Conference on Embedded and Ubiquitous Computing (EUC). IEEE, pp. 818–825 (2010)
16. Dempster, A.P.: A generalization of bayesian inference. In: Classic Works of the Dempster-Shafer Theory of Belief Functions, vol. 17. Springer, pp. 73–104 (2008)
17. USC/ISI, NS-2: The Network Simulator. http://www.isi.edu/nsnam/ns/

Representation and Validation of Enterprise Security Requirements—A Multigraph Model

Suvam Kr. Das and Aditya Bagchi

Abstract Starting with a formal model for detecting vulnerabilities and threats in an enterprise information system based on first-order logic, this research effort ventures to develop a multigraph model. First, it considers the physical interconnection graph that connects different hardware, software, and information resources present in the enterprise. It then generates a graph, based on different authorizations derived from managerial specifications. Superimposition of one graph over the other gives rise to the required multigraph. Using this multigraph, Erroneous specifications and Ambiguous specifications are identified. Vulnerable specifications are yet to be considered. Soundness and completeness of the methods have been discussed. Future research efforts related to the problem have also been indicated.

1 Introduction

An "enterprise" as defined in dictionary, is a business establishment involved in production or services. However for meeting the computational needs, a present-day enterprise may be considered as a large network of different resources. These resources, from the point of view of an establishment, cover all hardware equipment, software installed in them, applications executed, information stored, and the employees, usually acting as users of the computing resources. Considering these resources as assets to the enterprise, interrelationships among these assets may give rise to avenues of external attacks [1, 2]. In other words, these interrelationships may introduce inherent weaknesses or vulnerabilities that can be exploited as threats by

S. Kr. Das (✉)
Department of Computer Science, P.R. Thakur Government College, North 24 Pgs,
Thakurnagar 743287, West Bengal, India
e-mail: dassuvam88@gmail.com

A. Bagchi
School of Mathematical Sciences, Ramakrishna Mission Vivekananda University,
Belur Math, Howrah 711202, West Bengal, India
e-mail: bagchi.aditya@gmail.com

© Springer Nature Singapore Pte Ltd. 2018
R. Chaki et al. (eds.), *Advanced Computing and Systems
for Security*, Advances in Intelligent Systems and Computing 667,
https://doi.org/10.1007/978-981-10-8183-5_10

external agencies or even by some internal users. These vulnerabilities can be technical vulnerabilities which are listed and regularly updated in a database like ICAT Metabase of NIST and Mitre [3]. There are also several tools available for identifying technical vulnerabilities [4]. However, the Managerial policies as specified by the administrators/owners of an enterprise may introduce many undetectable vulnerabilities. An interesting effort has already been made to convert such managerial vulnerabilities to implementation-level vulnerabilities [5]. The same research group has also proposed a formal methodology to detect Managerial Vulnerabilities and Threats in an Enterprise Information System [6, 7]. Though there are other models available for security policy specification, the present research effort considers the model in [6, 7]. The primary reason is, this model is the continuation of a research effort that derives implementation-level authorization specification from higher level managerial specification [5] and ultimately arrives at an atomic authorization specification of the form (s_i, o_j, a_k, sg), where s_i is a subject that is authorized to access object o_j with access right a_k and sign $sg \in \{+, -\}$. A positive authorization allows an access and a negative one expresses explicit denial. In addition, the first-order logic-based specification formalism provides inference mechanism that helps in deriving new authorizations by inheritance from the authorizations explicitly specified. The present paper ventures to show that the formal specification made earlier in [6, 7] can be represented by a multigraph model so that an access request can be expressed as graph query, validated against the authorization specifications, and full or partial access may be granted.

The present paper covers a small portion of a much larger research effort. The total research scheme associated with the present paper will be discussed in the conclusion section detailing the future work. Concentrating on the present paper, we consider that any enterprise network can be modeled by one or more suitable graph structures. We start by specifying a Physical Interconnection Graph (PIG) that involves hardware components present in the Enterprise network connected by different semantic links/edges. The PIG is then augmented to place the software components in the graph including the applications with more semantic links. We then specify the access control policies imposed by the enterprise management in the form of another graph, namely the Policy graph. In this graph, all nodes may not be connected to each other and may have multiple isolated subgraphs. When the two graphs mentioned so far are superimposed on each other, a multigraph is created. When this multigraph is stored in a suitable graph database, any access to the Enterprise network becomes a query on the multigraph schema stored in the database. So this database can serve as a composite policy server for both positive and negative authorizations. This entire effort has been modeled in a single-user environment. This is the first stage of our work. The present paper covers this first stage to establish that

1. A multigraph model can be used to represent possible controlled access over an enterprise network detecting and removing any possible ambiguities and errors that may be present in managerial policy specifications.

2. The specification and representation are sound (free from all ambiguities and errors) and complete (all nodes of the multigraph are covered). In this respect, suitable lemmas and theorem have been proposed and proved to justify the completeness and soundness of specification.

The paper can be divided as follows: Sect. 2 discusses about the Physical Interconnection Graph (PIG). Section 3 provides the policy specification and multigraph. Section 4 discusses about the possible implementation of the multigraph structure using an open-source graph database system Neo4J. Section 5 draws the conclusion and discusses about future research efforts.

2 Physical Interconnection Graph (PIG)

Physical Interconnection Graph (PIG) contains the Hardware entities, Software entities including applications and the information storage like files and databases. The entire set of nodes and edges can be enumerated as entities and semantic links. Graphs in the proposed multigraph model have been defined as Property Graph [8], where a graph attaches properties to nodes and edges along with its interconnection structure. Well-known graph database system Neo4J [9] has also been developed on this property graph model.

In the property graph model, a graph can be specified in terms of the following definitions:

- N is a set of nodes, identified by an identifier $n.id$, $n \in N$.
- E is a set of directed edges, identified similarly by $e.id$, $e \in E$ with $N \cap E = \Phi$
- T, L: sets of node and edge labels respectively
- $\tau(n), N \to T$ is a function that assigns a node label to $n \in N$.
- $\rho(e), E \to L$ is a function that assigns an edge label to $e \in E$.
- A_N: a set of node attributes
- A_E: a set of edge attributes, where $A_N \cap A_E = \Phi$.

Note: If the nodes and edges have attributes of the same name, they will designate distinct attributes.

- $att_n: n \to A_j$, a function that assigns to node $n \in N$ a set of attributes $A_j \subset A_N$.
- $att_e: e \to A_k$, a function that assigns to edge $e \in E$ a set of attributes $A_k \subset A_E$.

However, for representing our multigraph model, a standard property graph needs to be extended to specify not only a graph structure but to impose necessary restrictions on access to different nodes and edges depending on access control policies. A graph constraint specification language (GCON) was developed in connection with Attack Graph specification [10] over a property graph. This earlier work has taken a database approach to store an attack graph in the open-source graph database Neo4J. Analysis of the attack graph can then be done by generating suitable queries on Neo4J.

We found that this design philosophy is very suitable for us. The Physical Inter-connection graph (PIG) with Hardware, Software and Information resources can be stored as a graph schema in the database. The second graph that specifies the access policies can then be imposed on the PIG exploiting the unmanaged service extensions facility of Neo4J to specify the access constraints.

2.1 Graph Constraint Specification Language for Multigraph Structure

Different types of nodes and their associated semantic interconnections have been detailed here. The constraints considered in these interconnections have also been explained. The specification mainly covers the nodes, semantic links, and associ-ated constraints required for PIG. Additional specifications required for a multiuser domain will be offered in future.

Label constraint A node may have the base labels from the set {*"Machine"*, *"OS"*, *"USB"*, *"ExtDrive"*, *"dataFile"*, *"SWPkg"*, *"DBMS"*, *"Disk"*, *"App"*}

set label from {*"Machine"*, *"OS"*, *"ExtDrive"*, *"dataFile"*, *"SWPkg"*, *"DBMS"*, *"Disk"*, *"App"*, *"USB"*}
An SWPkg node may have the additional label *"dataAccessService"*.
setExtraLabel ("SWPkg", "dataAccessService")

Uniqueness constraint Every node has a unique node identifier called "nid". *unique (n: n.nid)*
   There exists type-based unique identifier such as "OSID" or "DBMSID" for an OS or a DBMS node. *unique (n:OS, n.OSID), unique(n:DBMS, n.DBMSID)*

Degree constraint Some node labels (e.g., *ExtDrive*, *Disk*) will have no outgoing edges, i.e., the outdegree of these nodes must be 0.

set outdegree ((:ExtDrive)-[.]->()) = 0
set outdegree ((:Disk)-[.]->()) = 0
A *dataFile* can be *"locatedIn"* only one *disk*. Thus, there can be only one *"locatedIn"* edge from a *dataFile* to a *Disk*. *set degree ((:dataFile)-[:locatedIn]->(:Disk)) = 1*

Edge constraint The *"pConnect"* edge has the domain {*Machine*} and range {*Machine, Disk, ExtDrive*}

associate (Machine)-[pConnect]->(Machine)
associate (Machine)-[pConnect]->(Disk)
associate (Machine)-[pConnect]->(ExtDrive)

   The *"runsOn"* edge has domain {*DBMS, SWPkg, App*} and range {*OS*}.

associate (DBMS)-[runsOn]->(OS)

associate (SWPkg)-[runsOn]->(OS)
associate (App)-[runsOn]->(OS)

The *"executes"* edge has domain {*Machine*} and range {*OS*}

associate (DBMS)-[executes]->(OS)

The *"locatedIn"* edge has domain {*SWPkg, dataFile, OS, DBMS*} and range {*Disk*}

associate (SWPkg)-[locatedIn]->(Disk)
associate (dataFile)-[locatedIn]->(Disk)
associate (OS)-[locatedIn]->(Disk)
associate (DBMS)-[locatedIn]->(Disk)

The *"accesses"* edge has the following domains and ranges:

(a) From *dataAccessService* to *DBMS* provided they run on the same *OS* instance

associate (dataAccessService)-[accesses]-> (DBMS)

(b) From *SWPkg* to {*SWPkg, DBMS, dataFile*} provided they run on the same *OS* instance

associate (SWPkg)-[accesses]->(SWPkg)
associate (SWPkg)-[accesses]->(DBMS)
associate (SWPkg)-[accesses]->(dataFile)

(c) From *SWPkg* to *dataAccessService* provided they run on different *OS* instances

associate (SWPkg)-[accesses]->(dataAccessService)

(d) From *App* to {*SWPkg, DBMS, dataFile*} provided they run on the same *OS* instance

associate (App)-[accesses]->(SWPkg)
associate (App)-[accesses]->(DBMS)
associate (App)-[accesses]->(dataFile)

Participation and Cardinality Constraint Default participation of labels on edges is partial. Total-participation needs to be defined explicitly. Default cardinality of edges is many-to-many. Other cardinalities are to be defined explicitly. Since all the edges represented, here, can be considered as binary relations, we can associate them with cardinalities as per the description:

The *"runsOn"* edge is a total function, which has domain {*DBMS, SWPkg, App*} and range {*OS*}. That means for every instance of *DBMS, SWPkg,* and *App*, there must exist an instance of *OS*. This depicts that participation of {*DBMS, SWPkg, App*} on *runsOn* is total and the *runsOn* edge has cardinality many-to-one.

set total-participation of "DBMS" on "runsOn"
set total-participation of "SWPkg" on "runsOn"

set total-participation of "App" on "runsOn"
set cardinality("runsOn", many-to-one)

The *"executes"* edge is a total function from {*Machine*} to {*OS*}. Hence, every machine must execute one *OS* instance (which must reside in a *Disk* of the same machine) at a time. One *OS* instance is also executed in only one machine.

This depicts that participation of {*Machine*} on *executes* is total and *executes* edge has cardinality one-to-one.

set total-participation of "Machine" on "executes"
set cardinality("executes", one-to-one)

The *"locatedIn"* edge is a total function and has domain {*SWPkg, dataFile, OS, DBMS*} and range {*Disk*}, provided the disk is on the same machine.

set total-participation of "SWPkg" on "locatedIn"
set total-participation of "dataFile" on "locatedIn"
set total-participation of "OS" on "locatedIn"
set total-participation of "DBMS" on "locatedIn"
set total-participation of "Disk" on "locatedIn"

The *"accesses"* edge represents a partial function with domain {*dataAccessService, SWPkg, App*} and range {*SWPkg, DBMS, dataFile, dataAccessService*}

As per the definition, the participation and cardinality have the default values.

A sample PIG based on the above specifications is shown in Fig. 1. It shows a set of hardware and software resources connected to each other by the semantic links available in the specification language. For example, *Machine* (M2) is physically connected to *Disk* (Disk3) through *pConnect* edge and M_2 *hasOS* OS2 which is *locatedIn* Disk3.

3   Policy Specification and Multigraph

As mentioned earlier, an atomic authorization specification is of the form (s_i, o_j, a_k, sg), where s_i is a subject, that is, authorized to access object o_j with access right a_k and sign sg ϵ {+, −}. A positive authorization allows an access and a negative one expresses explicit denial.

Objects, as considered in the enterprise information system, will be available from the Physical Interconnection Graph (PIG). These objects are the hardware assets, software assets, and information assets. Using the physical interconnections in PIG, two relationships among the objects, namely *inclusion* and *connection* have been defined for policy specification and to derive new policies by inheritance from the explicitly specified policies. In other words, these two relationships help in deriving implicit authorizations from the authorizations explicitly specified.

Fig. 1 A sample physical
interconnection graph

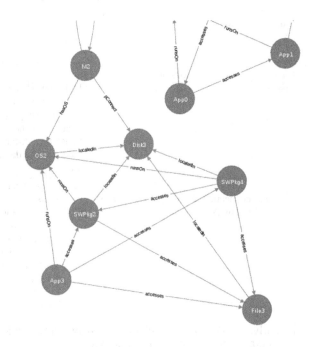

Incidentally, this paper has considered a single-user model. So there is no conflict among users through user-group or role hierarchy. Conflicts that may arise between explicit and implicit authorizations are only due to erroneous or ambiguous specifications made by the enterprise management. The purpose of the multigraph model and derivation of policies within it, is to show that the proposed model is in a position to identify the erroneous and/or ambiguous policy specifications. Without interpreting them as vulnerabilities, enterprise management is reported with the errors to rectify managerial specifications.

Table 1 shows the properties of the two relationships *inclusion* and *connection,* which clarify why they can be used for deriving new policies from the policies explicitly specified.

Let us discuss with examples, how *inclusion* and *connection* relationships help in deriving new policies. For example, an explicit access right given to a database infers access to the hardware platform and also to the operating system of the server where the database is stored even when they are not explicitly specified. It gives rise to a series of *inclusion* relationships as

$$\text{DBMS} \; \gamma \, \text{OS} \; \gamma \, \text{HW}$$

So, from a positive authorization given to a user u_i to access DBMS will infer other authorizations as

$$(u_i, \text{DBMS}, a_k, +) \Rightarrow (u_i, \text{OS}, a_k, +) \Rightarrow (u_i, \text{HW}, a_k, +)$$

Table 1 Relationships among the objects

Relation (Θ)	Properties
Inclusion (γ) The set of inclusion relations of an enterprise is represented by Γ	$o_x \gamma o_y \Rightarrow$ object o_x is included in object o_y $o_x \gamma o_y \not\Rightarrow o_y \gamma o_x$ Inclusion relation is noncommutative $o_x \gamma o_y \wedge o_y \gamma o_z \Rightarrow o_x y o_z$ Inclusion supports transitivity. Since inclusion relation is noncommutative and supports transitivity, different objects related by inclusion can define a hierarchy (i.e., *object hierarchy*)
Connection (δ) The set of connection relations of an enterprise is represented by Δ	$o_x \delta o_y \Rightarrow$ object o_x is connected to object o_y $o_x \delta o_y \Rightarrow o_y \delta o_x$ Connection relation is commutative $o_x \delta o_y \wedge o_y \delta o_z \Rightarrow o_x \delta o_z$ Connection supports transitivity. Since connection relation is commutative and supports transitivity, different objects related by connection can define a *connection chain*

Similarly, for a *connection* relationship between two objects o_i and o_j, access right to one infers access right to the other. So,

$$(o_i \, \delta \, o_j) \wedge (u_i, o_i, a_k, +) \Rightarrow (u_i, o_j, a_k, +)$$

Reachability (R): Combination of *inclusion* and *connection* offers the notion of *reachability*. A *reachability* function $R(o_i, o_j)$ = True implies that object o_j is reachable from o_i using inter-object relationship.

So, if a digraph is drawn using objects as nodes and *inclusion* and *connection* relationships as edges, as obtained from the Physical Interconnection graph (PIG, $R(o_i, o_j)$ = True implies that in the PIG, there is a path from node o_i to node o_j.

It is assumed that for accessing any of the assets (objects), a user must have either an explicit positive authorization or an inferred implicit positive authorization for the object. A managerial specification may also be derived as an explicit denial. Because of the presence of both positive and negative authorizations and explicit and implicit or inferred authorizations, access to a node from another node may be undecidable. In other words, a user accessing an object may infer that it has both positive and negative authorizations to it for the same access right. The authorizations as obtained from the managerial specifications may have some ambiguities wherefrom such undecidable situation may arise. Before detection of such ambiguities, next section studies the policy inference mechanism.

Authorization: An authorization is usually specified as (s_i, o_j, a_k, sg), where s_i is a subject that is authorized to access object o_j with access right a_k and $sg \in \{+, -\}$. A positive authorization admits an access and a negative one expresses explicit denial.

This single authorization may be considered as an atom in the policy set of the enterprise.

Policy: A policy is either an atom (a single authorization) or a number of authorizations connected by Boolean operators And, Or, Exclusive-Or, etc. A policy set is formed by number of policies separated by commas.

As mentioned earlier, in this paper only object-to-object interconnection and hierarchy have been considered. There can also be user hierarchies, where user-groups may be formed and each individual user may be a member of one or more groups. Any authorization given to a user-group will be inherited by all its members. Moreover, a user-group can be the sub-group of another larger user-group and the sub-group will inherit the access authorizations of its super-group [11]. There can also be role hierarchies among users and these users are assigned to different roles giving ways to authorization inheritance again [12]. Such studies will be made later. For the present paper, the study is restricted to the detection of authorization specification related ambiguities arising out of object interconnection and hierarchy only. So, similar to PIG, a policy graph can be generated for each user where each authorization will link a user/subject to an object starting from the objects or hardware platforms where the concerned user has valid usercodes. For this paper, it has been assumed that validation of managerial policy specifications after deriving to lower level implementable specification has been done for each individual user.

Wikipedia defines a multigraph as

"Multigraph is a graph which is permitted to have multiple edges (also called parallel edges), that is, edges that have the same end nodes. Thus two vertices may be connected by more than one edge." Because of the superimposition of policies over the physical interconnection of objects, two nodes may be connected by more than one edge:

Figure 2 shows a sample multigraph, where user u_i has a valid usercode in object o_i. So, u_i is connected to o_i by *inclusion* relationship. Now, deriving from managerial policy specification, if u_i gets an authorization $(u_i, o_j, a_k, +)$ but u_i does not have a usercode in o_j, then the access will be granted only if o_j is reachable from o_i, i.e., $R(o_i, o_j) = \text{True}$. For similarity of representation let us consider that $R(o_i, o_j, +)$ represents o_j is reachable from o_i, else it is $R(o_i, o_j, -)$. So, in this example u_i will access o_j if,

$$R\left(o_i, o_j, +\right) \wedge \left(o_i, o_j, a_k, +\right) \Rightarrow \left(u_i, o_j, a_k, +\right)$$

Figure 2 shows a direct *connection* between o_i and o_j, however in practice if o_j is reachable from o_i in the Physical Interconnection Graph, u_i will be able to access o_j.

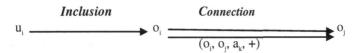

Fig. 2 A sample multigraph

The present model has considered superimposition or interplay between only two graphs: Physical Interconnection Graph and the Policy Graph. In addition, the model in its full form will also consider User-group hierarchy as a graph and policy inheritance therein. User query over this multigraph should also be a graph which may be fully or partially served depending on the accesses permitted in the query against the physical connection and policy specified or derived. However in order to highlight the possible ambiguities in managerial policy specifications, this paper considers only the two graphs mentioned above.

While deriving lower level specifications from managerial policies, many ambiguities may arise. General practice is to search for possible vulnerabilities. However, introduction of Physical Interconnection Graph helps in discarding ambiguous authorizations and report them to enterprise administration for corrective actions. These errors and ambiguities have been classified as follows.

Erroneous Specification: A positive authorization specified as $(u_i, o_j, a_k, +)$ is considered erroneous, if u_i accesses o_j through o_i, i.e., u_i has a usercode in o_i but PIG shows $R(o_i, o_j, -)$, i.e., o_j is not reachable from o_i. In other words, authorization has been given to an object for accessing another object but the two objects are not reachable from each other on the Physical Interconnection Graph.

Another type of erroneous specification can be made if two conflicting authorizations are specified in an *inclusion* path. Considering the DBMS access mentioned earlier, two authorizations are specified as

$$(u_i, DBMS, a_k, +) \wedge (u_i, HW, a_k, -)$$

Here, DBMS is running in the hardware platform HW. User u_i has a positive authorization for accessing the DBMS, but it also has an explicit denial for accessing the hardware platform where DBMS is situated. An erroneous authorization can be detected at the specification stage itself by placing it over the PIG, without storing such authorizations in the policy store.

Ambiguous Specification: Example of ambiguous specification is shown in Fig. 3.

Figure 3 shows an ambiguous specification. A user u_i may have usercode in both o_i and o_k. However, accessing object o_j from them gets conflicting authorizations and makes the access undecidable. This problem of undecidability has been studied by many researchers starting from the seminal paper of Harrison, Ruzzo, and Ullman (HRU) [13]. Policy specification is considered to be sound if for any user accessing any object with any access right, no undecidable situation is encountered. In some restrictive cases problem of safety and undecidability have been solved, but for any arbitrary Physical Interconnection Graph with both positive and negative authorizations, it is yet to be achieved. However, the present approach verifies the authorization in the specification stage itself without storing it in the policy store. So in case of any

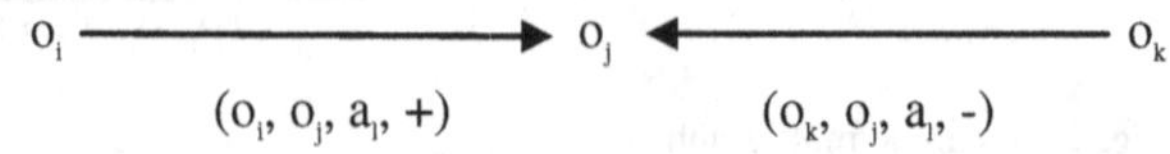

Fig. 3 An ambiguous specification

$$o_i \longrightarrow o_j \longleftarrow o_k$$

$$(o_i, o_j, a_l, +) \qquad (o_k, o_j, a_l, -)$$

ambiguous specification, both the positive and negative authorizations are discarded and the enterprise administration is reported for corrective actions. However, if this undecidable situation occurs between an explicit authorization and an implicit authorization (authorization derived from *Inclusion* and *Connection* links), the explicit authorization will get preference. Justification for this rule will be discussed later. One may always argue that the ambiguous specification defined here is actually identifying some inconsistencies in the specification and should also be categorized as erroneous specification again. However, the authors believe that the erroneous specifications are those which cannot be implemented at all, whereas later category creates an undecidable situation. It will be even more apparent when user hierarchy is added to the system either as a user-group hierarchy and/or role hierarchy.

Vulnerable specification: Security of access of any object in an enterprise information system is usually defined by three security parameters: Confidentiality, Integrity, and Availability. While Confidentiality controls access to an object, Integrity ensures that any data can be altered only by authorized users. Availability ensures that an authorized user should get access to objects assigned to him/her. The Formal model considered for the creation of the present multigraph model has shown that vulnerability of any object for all the three security parameters can be checked in linear time. However, if it is to be done for all the objects, the time complexity of the algorithm is of the order of n^2 [6, 7]. Study of the vulnerability of specification will be done in future and not within the scope of this paper.

However, the present paper should meet completeness and soundness criteria with respect to the detection of erroneous and ambiguous specifications. Considering the discussion so far, authorization specification should adopt the following authorization types:

Definition (Explicit Authorization): An explicit authorization is a lower level implementable authorization directly derived from the managerial specification.

Definition (Implicit Authorization): An implicit authorization is a lower level implementable authorization derived/inferred from an explicit authorization by the inference mechanism of *Inclusion* and/or *Connection* relations present among nodes in PIG.

Definition (Default Authorization): It is necessary for an enterprise that for each user and object combination a definite authorization can be inferred. It is the completeness criterion. However, even with the specification of both positive and negative authorizations, it cannot be ensured. In the real-life situation, authorizations are specified for only a few object-user combinations. Inference mechanism may not meet the completeness criterion. As a result, against a query, a policy store may not be able to decide whether a particular user can access a particular object or not. In other words, there may not be any authorization specified for such object-user pair, either positive or negative. It is also not possible, to specify explicit authorizations for any object-subject/user pair, since all users may not be able to reach all objects. So, in order to ensure completeness, two default authorization specification policies are employed: *Close policy*, where all object-user combinations are assumed to have negative authorizations and access may be permitted by explicit positive authorizations and *Open policy*, where all object-user combinations are assumed to have positive authorizations and access may be blocked by explicit negative authorizations [9]. Each subnet within the enterprise may declare either close policy or open policy for a user or a user-group.

Lemma 1 *Employing close and open policies, completeness of policy specification can be ensured.*

Proof

1. A close or an open policy will ensure that all objects under such environment will by default offer, respectively, a negative access or a positive access to all users for all types of access rights.
2. Since any user will encounter either a close or an open policy against any object, it will have either a negative access or a positive access by default.
3. Managerial specifications will actually specify exceptions to such close or open policies by offering explicit positive or negative authorizations.

Hence no query will ever encounter a situation where any object-subject/user combination will have no authorization assigned.

This situation, however, does not ensure that there will be no erroneous or ambiguous authorization specification.

Soundness Criterion: Soundness criterion must ensure that for an object-user combination and a particular access right, either a positive or a negative authorization would be encountered but never both. In practice, various policies have been adopted to solve this undecidable situation. Some may give preference to negative authorization suspending the positive one and some may give preference to explicit authorization over derived/implicit authorization. The authors of the present paper prefer to suspend both the conflicting authorizations and to report to the administrator for resolving the problem. Meanwhile, any query related to affected object-user combinations would also remain suspended.

So, discarding such ambiguous specifications, rest of the policy set should offer sound authorization. As mentioned earlier, an authorization specification can be explicit, implicit or default. A default authorization can again be obtained either from a *Close* or an *Open* policy. An implicit authorization will be derived from the combination of Physical Interconnection Graph and Policy Graph. Let P be the set of policies covering explicit authorizations only. P_{der} is the set of derived policies covering default and other implicit authorizations where an authorization or an atomic policy is represented by the tuple (s_i, o_j, a_k, sg), as mentioned earlier. Authorization specification, derivation, and validation processes can be summarized by the following rule set:

[Rule-1] *Reflexivity Rule*: All tuples of P are inherited by P_{der} after the removal of erroneous and ambiguous managerial specifications.

[Rule-2a] *Inheritance Rule (Inclusion)*: $(o_x \; \gamma \; o_y) \wedge (u_i, o_x, a_k, +) \Rightarrow (u_i, o_y, a_k, +)$, where $(o_x \; \gamma \; o_y) \in$ PIG, $(u_i, o_x, a_k, +) \in P$ and $(u_i, o_y, a_k, +) \in P_{der}$.

Note: While the object o_x is included in object o_y in the Physical Interconnection Graph (PIG) and a user u_i has a positive authorization on o_x for an access right a_k, then the user u_i should get a positive authorization on o_y in order to make the access to object o_x possible.

Erroneous Specification: If $(o_x \ \gamma \ o_y) \wedge (u_i, o_x, a_k, +) \wedge (u_i, o_y, a_k, -) \Rightarrow$ Error, where

$(o_x \ \gamma \ o_y) \in$ PIG, $(u_i, o_x, a_k, +) \in P$ and $(u_i, o_y, a_k, -) \in P$.

[Rule-2b] *Inheritance Rule (Connection)*: $(o_x \ \delta \ o_y) \wedge (u_i, o_x, a_k, +) \Rightarrow (u_i, o_y, a_k, +)$, where $(o_x \ \delta \ o_y) \in$ PIG, $(u_i, o_x, a_k, +) \in P$ and $(u_i, o_y, a_k, +) \in P_{der}$.

Note: While the object o_x has a connection with the object o_y in PIG and a user u_i has a positive authorization on o_x for an access right a_k, then the user u_i will infer a positive authorization on o_y as well unless there is an explicit negative authorization specified.

Ambiguous Specification: $(o_x \ \delta \ o_y) \wedge (u_i, o_x, a_k, +) \Rightarrow (u_i, o_y, a_k, +)$ and $(o_z \ \delta \ o_y) \wedge (u_i, o_z, a_k, -) \Rightarrow (u_i, o_y, a_k, -)$. So by Rule-2b P_{der} will inherit, $(u_i, o_y, a_k, +) \wedge (u_i, o_y, a_k, -) \Rightarrow$ Error.

[Rule-3a] *Override Rule (between Explicit and Default authorization)*: If $(u_i, o_x, a_k, +) \in P$ and $(u_i, o_x, a_k, -) \in P_{der}$ (Default authorization by *Close* policy rule) then $(u_i, o_x, a_k, +) \in P_{der}$.

Similarly, if $(u_i, o_x, a_k, -) \in P$ and $(u_i, o_x, a_k, +) \in P_{der}$ (Default authorization by *Open* policy rule) then $(u_i, o_x, a_k, -) \in P_{der}$.

Note: Any Explicit authorization specification overrides a Default authorization specified by a *Close* or *Open* policy.

[Rule-3a] *Override Rule (between Implicit and Default authorization)*: $((o_x \ \gamma \ o_y) \vee (o_x \ \delta \ o_y)) \wedge (u_i, o_x, a_k, +) \Rightarrow (u_i, o_y, a_k, +)$. Then according to Rule-2a and Rule-2b, $(u_i, o_y, a_k, +) \in P_{der}$. Now if there already exists $(u_i, o_y, a_k, -) \in P_{der}$ as a Default authorization because of a *Close* policy, then Implicit authorization will override the Default authorization and authorization $(u_i, o_y, a_k, +) \in P_{der}$ will persist.

Similar situation will occur when an Implicit negative authorization will be in conflict with a Default positive authorization for a *Open* policy and the Implicit negative authorization will be included in P_{der} replacing the Default authorization.

Lemma 2 *Under the given rule set, all authorization specifications are sound.*

Proof All authorization specifications are sound if for any object-user combination and for any access right there is no undecidable situation. Access authorization to an object for a user-access right combination will be undecidable if the concerned user for a particular access right infers both positive and negative authorizations to the object. First, such situation may occur if two Explicit authorizations or Implicit authorizations are in conflict. Second, if an Explicit or Implicit authorization is in conflict with a Default authorization.

1. If both positive and negative authorizations are explicit then an "Error" condition is generated and both the authorizations are discarded for possible change/correction in corresponding managerial specification **[Rule-2a]**.
2. If both positive and negative authorizations are inferred from explicit authorizations, i.e., Implicit authorizations, then also an "Error" condition is generated

and both the authorizations are discarded for possible change/correction in corresponding managerial specification **[Rule-2b]**.

3. If explicit/implicit positive authorization is in conflict with a default negative authorization offered by a prevailing *Close* policy or if explicit/implicit negative authorization is in conflict with a default positive authorization offered by a prevailing *Open* policy, the explicit/implicit authorization is retained and the default authorization is discarded **[Rule-3a and Rule 3b]**.

Since no other undecidable authorization specification is possible, after removing the "Error" condition, P_{der} will include all authorization tuples of P and all such authorizations will be sound.

Theorem 1 *Authorizations specified over the multigraph generated by the superimposition of Physical Interconnection Graph and Policy graph are sound and complete.*

Proof Proof is directly derivable from Lemma 1 and Lemma 2.

4 Realization of Multigraph and Experimental Setup

For the purpose of implementation, Neo4J graph database has been chosen to map the Physical Interconnection graph to its Property Graph Model. Neo4J supports unmanaged service extensions to specify constraints. Authorizations have been modeled this way. Verification of access right to any object for any user can be done by generating an appropriate query on the database. However, this entire effort is for the proposed single-user model. Our approach for multiuser situation particularly under user hierarchies may be different and yet to be studied thoroughly.

5 Conclusion and Future Work

The present paper being the first attempt to report a research effort, most of the planned work will be referred to future work. So far, only Erroneous and Ambiguous specifications have been studied. Vulnerability analysis will be attempted with multiple users placed in user-group and/or role hierarchies by augmenting the multigraph structure with a third graph representing user hierarchy. An important work has been planned about restructuring of network. Reallocation of resources is a common method for resolving vulnerabilities and for mitigating threats. However, such reallocation may introduce new vulnerabilities even worse than the original one. Since the resources to be secured are usually classified as low-, medium- or high-risk components, a computational model can be attempted to quantify the total risk involved in any reallocation of resources. It is believed, that an optimization function may be obtained to compare among different reallocation strategies so that ultimately against each change in the network or policy set, a resource distribution plan for minimum risk may be obtained.

The vulnerability and threat that may occur for any change in the physical network or in policy specification will also be covered in future.

References

1. Information Technology—code of practice for information security management. ISO/IEC 17799:2005
2. IT baseline protection manual. BSI, 2004
3. ICAT metabase—http://cve.mitre.org
4. Top 10 vulnerability scanners—http://sectools.org/vuln-scanners.html
5. Sengupta, A., Mazumdar, C., Bagchi, A.: A methodology for conversion of enterprise-level information security policies to implementation-level policies/rules. In: Proceedings of the 2nd International Conference on Emerging Applications of Information Technology (EAIT 2011), pp. 280–283. IEEE Press, New York, February (2011). https://doi.org/10.1109/eait.2011.87
6. Sengupta, A., Mazumdar, C., Bagchi, A.: A formal methodology for detection of vulnerabilities in an enterprise information system. In: Proceedings of 4th International Conference on Risks and Security of Internet and Systems (CRISIS 2009) pp. 74–81. October (2009). https://doi.org/10.1109/crisis.2009.5411976
7. Sengupta, A., Mazumdar, C., Bagchi, A.: A formal methodology for detecting managerial vulnerabilities and threats in an enterprise information system. Journal of Network and Systems Management 19(3), 319–342 (2011). https://doi.org/10.1007/s10922-010-9180-y
8. Property graph model. https://github.com/tinkerpop/blueprints/wiki/Property-Graph-Model
9. Neo4J graph database. https://neo4j.com/
10. Barik, M.S., Mazumdar, C., Gupta, A.: Network vulnerability analysis using a constrained graph data model. In: Proceedings 12th International Conference, ICISS 2016. LNCS, vol. 10063, pp. 263–282 (2016)
11. Jajodia, S., Samarati, P., Subrahmanian, V. S.: A logical language for expressing authorizations. In: Proceedings of the IEEE Symposium on Security and Privacy, pp. 31–42. Oakland, California, USA (1997)
12. Sandhu, R.S., Coyne, E.J., Feinstein, H.L., Youman, C.E.: Role-based access control models. IEEE. Computer 29(2), 38–47 (1996)
13. Harrison, M.A., Ruzzo, W.L., Ullman, J.D.: On protection in operating systems. In: Proceedings of Fifth Symposium on Operating Systems Principles, pp. 14–24. The University of Texas at Austin, USA (1975)

Author Index

© Springer Nature Singapore Pte Ltd. 2018
R. Chaki et al. (eds.), *Advanced Computing and Systems
for Security*, Advances in Intelligent Systems and Computing 667,
https://doi.org/10.1007/978-981-10-8183-5

Printed in the United States
By Bookmasters